Won H. Kim

Science of New Paradigm

AF153314

Won H. Kim

Science of New Paradigm

3D Wave: New Paradigm Beyond Materialism

LAP LAMBERT Academic Publishing

Impressum / Imprint
Bibliografische Information der Deutschen Nationalbibliothek: Die Deutsche Nationalbibliothek verzeichnet diese Publikation in der Deutschen Nationalbibliografie; detaillierte bibliografische Daten sind im Internet über http://dnb.d-nb.de abrufbar.
Alle in diesem Buch genannten Marken und Produktnamen unterliegen warenzeichen-, marken- oder patentrechtlichem Schutz bzw. sind Warenzeichen oder eingetragene Warenzeichen der jeweiligen Inhaber. Die Wiedergabe von Marken, Produktnamen, Gebrauchsnamen, Handelsnamen, Warenbezeichnungen u.s.w. in diesem Werk berechtigt auch ohne besondere Kennzeichnung nicht zu der Annahme, dass solche Namen im Sinne der Warenzeichen- und Markenschutzgesetzgebung als frei zu betrachten wären und daher von jedermann benutzt werden dürften.

Bibliographic information published by the Deutsche Nationalbibliothek: The Deutsche Nationalbibliothek lists this publication in the Deutsche Nationalbibliografie; detailed bibliographic data are available in the Internet at http://dnb.d-nb.de.
Any brand names and product names mentioned in this book are subject to trademark, brand or patent protection and are trademarks or registered trademarks of their respective holders. The use of brand names, product names, common names, trade names, product descriptions etc. even without a particular marking in this works is in no way to be construed to mean that such names may be regarded as unrestricted in respect of trademark and brand protection legislation and could thus be used by anyone.

Coverbild / Cover image: www.ingimage.com

Verlag / Publisher:
LAP LAMBERT Academic Publishing
ist ein Imprint der / is a trademark of
OmniScriptum GmbH & Co. KG
Heinrich-Böcking-Str. 6-8, 66121 Saarbrücken, Deutschland / Germany
Email: info@lap-publishing.com

Herstellung: siehe letzte Seite /
Printed at: see last page
ISBN: 978-3-659-53713-4

Copyright © 2014 OmniScriptum GmbH & Co. KG
Alle Rechte vorbehalten. / All rights reserved. Saarbrücken 2014

Table of contents

Part I. Science of New Paradigm

Part II. New Approach Controlling Cancer: Water Memory

1 Introduction

2 Method

3 Results

4 Discussion

Part I

Science of New Paradigm

Science pursues truth. However, scientific quest for truth is confined in the frame of a paradigm. Gödel mathematically proved that for any system, there will always be statements that are true, but that are unprovable within the system. There always appear facts which cannot be explained with current paradigm. Such fact can be judged only at higher dimension. Paradigm shift is unavoidable.

Paradigm shift is initiated from unexplainable fact with current paradigm. Quantum science came from the trial to explain the unexplainable fact, black body radiation. Science of 21st century needs new paradigm. New paradigm will be initiated from the unexplainable fact. This time it could be water memory.

Science of new paradigm will be introduced in this book. New paradigm explains water memory with 3 dimensional wave (3D wave) inherent to matter. 3D wave is faster than speed of light and exists in the area of imaginary number. 3D wave can be separated from the matter to water. 3D wave in water as separate entity can be interacting with other matter.

Beyond analogue type of water memory 3D wave can be digitized and expressed in 2D space. Digitized 3D wave shows the same functionality of original matter. 21st century should be an era for new paradigm beyond materialism.

1. Science of new paradigm needed to explain water memory

Science pursues truth. However, scientific quest for truth is confined in the frame of a paradigm. Gödel mathematically proved that for any system, there will always be statements that are true, but that are unprovable within the system (incompleteness theorems) [1].

There always appear facts which cannot be explained with current paradigm. Such fact can be judged only at higher dimension. Paradigm shift is thus unavoidable.

Paradigm shift is initiated from the unexplainable fact with current paradigm. Quantum science came from the trial to explain the unexplainable fact, black body radiation. Quantum science began by assuming energy is not continuous but exist as a tiny discrete packet of which the minimal unit is hυ. Although quantum science showed wave part of the matter exist, modern physics focus only on particle. The paradigm of current science is based on particle science, which is materialism.

21st century needs new paradigm. New paradigm will be initiated from the unexplainable fact. This time it could be water memory which is not explained with current science based on materialism.

2. Homeopathy and water memory

The story of water memory begins from homeopathy. Homeopathy views diseases symptoms originated from the human body's natural healing process. Homeopathy uses toxic substance to enhance the ability of natural healing. It is thus very natural that homeopathy tried dilution method to reduce harmful effect of toxin. Homeopathic dilution is carried out with physical stimulation (vigorously shaken by 10 hard strokes at each dilution termed as succussion in homeopathy). What homeopathy found is that even though the poisonous substance gets diluted to the point where molecule is no longer exist, its effect still remained, which should mean 'water memory'.

Even though it has been used successfully clinically for over 200 years, homeopathy has been ignored by the orthodox medical circles until now because current science could not explain how a substance's effect to be shown without actual substance. However, there have been almost three hundred verification experiments done on homeopathic effects in the past decade. Furthermore, about 80% of those showed that homeopathy have different effects from placebo effect [2].

If homeopathy is effective therapy, it implies that water could store the information of toxins to boost natural healing power. If water could store the information of material, then this capacity does not need to be confined to homeopathy which uses mostly toxin.

3. Scientific approach begins from Benveniste

Although French scientist Benveniste did not know about homeopathy, he was the first to show water's memorizing ability in scientific way. In 1988 Benveniste and colleagues published a controversial article showing a biological reaction of ultra-high diluted solution of antiserum against IgE [3].

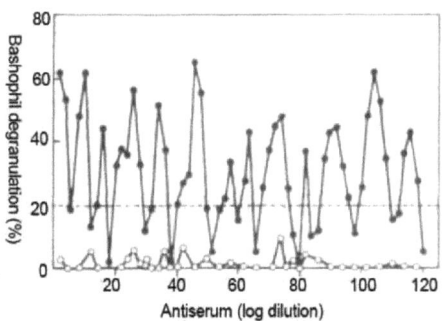

Basophil is degranulated by binding of its antiserum to surface and degranulated basophil releases histamine. Basophil is colored blue by the dye called toluidine blue. However, when histamine is released, basophil loses blue color.

In the paper it was demonstrated that human basophil degranulation was triggered by extremely diluted solution (10^{-120} of its original density) of antiserum where there left no molecule, which could be called as 'water memory'. This phenomenon was observed only when it is violently shaken at each dilution like homeopathy. Another interesting thing is basophil degranulation showed a phase of rise-fall repetition in the shape of a wave depending on the degree of dilution.

At first Benveniste could not believe what he found. Through repeated experiment he could convince his results, and invited three other laboratories worldwide, in Canada, Italy, and Israel to join his research. After they all get to the same conclusion through repeated experiments, they decided to publish in *Nature*.

Although *Nature* allowed publication, a follow-up investigation team was sent to Benveniste's lab, and they failed to replicate the original results (3 of 4 initial attempts turned out favorable to Benveniste, however 3 more trials after they designed a double blind procedure turned out negative). Benveniste refused to retract his controversial article, explaining that the protocol used in their trial was not identical to him [4]. However, his reputation was damaged and external source of funding was withdrawn. He continued his researches, and published many papers showing water memory effect under various experimental conditions [5-7].

4. Many investigations followed with various seriousness

As biological reaction in the absence of any effective molecules cannot be explained by conventional theory focusing only on the water structure of which the hydrogen bonds last only picoseconds scale, the results of

Benveniste and colleagues sparked many investigations with variety of seriousness.

The most serious one was the research performed double blind by 4 independent European laboratories in 2004 [8]. The interesting fact was the purpose of their double blind test was to disprove Benvensite's research but instead proved that it was in fact true.

One of participating researcher Ennis led the activities at the British lab. Ennis states that she began the research as a skeptic, but concluded that the "Results compelled me to suspend my disbelief and start searching for rational explanations for our findings" [9].

5. Review of water memory after 20 years

In 2010 a review of the attempts to replicate studies into the activation and inhibition of human basophils with homeopathic dilutions by Ennis was published in the journal *Homeopathy* [10]. The paper reviewed a list of studies to find out what can be confidently said about the 20 years of research into the subject.

Ennis concluded that there appears to be some evidences for an effect and it is needed to investigate how many of the effects are due to artifacts. She believes that in order to draw the conclusion of homeopathic inhibition of basophil a new multi-center trial would be required. However, all the attempts so far are poorly standardized between laboratories. Most important thing is before such a trial there should be agreement about how to undertake the experiment. Only such an approach might provide a definitive result.

6. New approach controlling cancer: water memory

Water memory could be reproduced in my laboratory. P53 is a DNA binding protein and functions as a potent tumor suppressor. However, there is virtually no practical way to utilize the function of P53 clinically. If the molecular structural information of P53 (this wave part of the matter is expressed as '3D wave' throughout this book, as it decreases thermodynamic entropy and maintains 3D structure unlike spreading electromagnetic wave) could be transferred to water or any medium contacting water, various strategies could be possible.

New electronic device was devised to transfer 3D wave of the matter using 7.8 Hz, resonance frequency of earth, instead of time consuming homeopathic technique. Subtle magnetic field generated by 7.8 Hz frequency could activate and transfer the 3D wave of the matter to water and other medium.

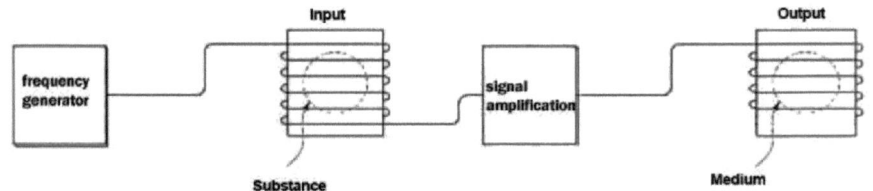

Block diagram of 3D wave transferring machine

The water to which 3D wave of P53 was transferred actually inhibited cancer proliferation, showed anti-metastasis, and increased apoptosis. The results have been published at the first issue of *Journal of Vortex Science and Technology* [11] and explained in detail at the part II of this book.

7. 3D wave of DNA imprinted to water

Nobel laureate Montagnier and colleagues showed that DNA polymerase could recognize 3D wave of specific DNA sequence of HIV transferred to water and produce new DNA copies [12] using PCR (Polymerase Chain

Reaction). PCR is a general technique which amplifies specific DNA. PCR needs specific DNA material for original copy.

They could amplify specific HIV DNA out of pure water to which 3D wave of HIV DNA was transferred. This result suggests that 3D wave of DNA is the physical entity recognized by DNA polymerase, even though it is not a visible (detectable) particle. They could detect electromagnetic signal from the water to which 3D wave of DNA was transferred. Montagnier and colleagues also used similar device to ours using 7.8 Hz frequency to transfer the 3D wave of DNA to water.

8. Phantom DNA

The existence of 3D wave of DNA as physical entity in space was also shown in so called 'phantom DNA effect' by Poponin [13]. He showed that scattering pattern of DNA by laser radiation could be regenerated even after DNA sample was removed.

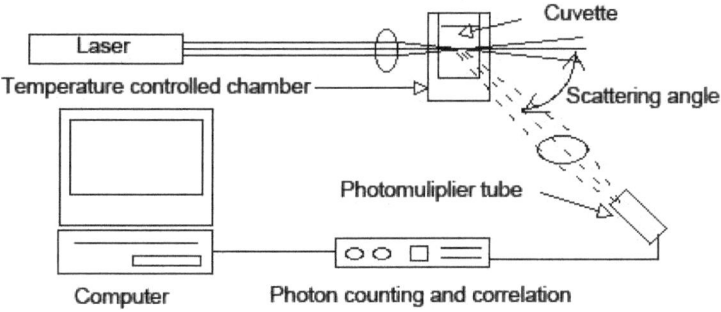

Block diagram of the laser photon correlation spectrometer

When the scattering chamber is void of physical DNA, scattered light showed only background random noise of the photomultiplier. When a physical DNA sample is placed in the scattering chamber, scattered light showed an oscillatory and slowly exponentially decaying pattern. When the

11

DNA is removed from the scattering chamber, it is expected that it will be the same as before the DNA was placed in the scattering chamber. Surprisingly scattered light pattern appeared distinctly different from the one obtained before the DNA was placed in the chamber, which is rather similar to that of DNA material although its intensity is about 10,000 times smaller.

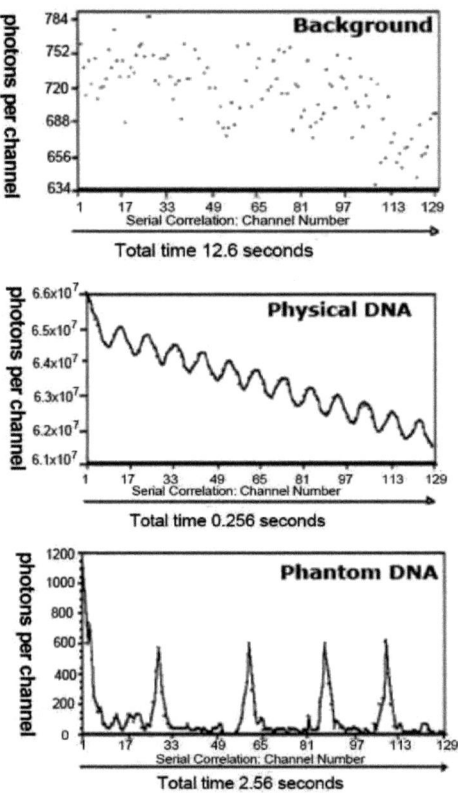

DNA phantom effect

As DNA scattering pattern appears without presence of DNA, this phenomenon is called as 'phantom DNA effect'. 'Phantom DNA effect' could be regenerated until about a month after DNA was removed. 'Phantom DNA effect' as well as other phantom phenomena such as 'phantom leaf' and 'phantom limb' suggests that there is an entity of the matter other than physical matter existing as a field with certain 3D structure.

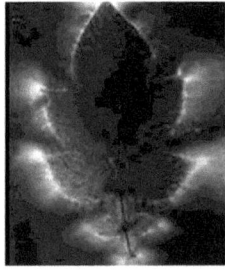 

Phantom Leaf: Kirlian photography of leaf showing missing part as 3D field

9. Inherent wave of the matter by de Broglie and Bohm

Benveniste thought that there is a wave inherent to molecule. When the wave propagated from molecule is transferred to a cell receptor through water, the wave can induce resonance of a receptor initiating intracellular signal transmission [5-7].

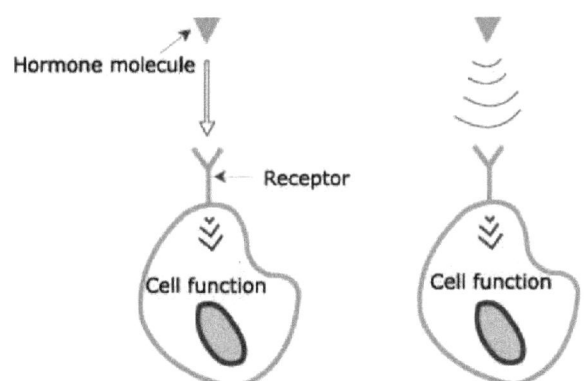

Comparison of current theoy(left) and Benvenist's Digital Biology(reight)

Inherent wave of matter is not a new concept. In 1924 de Broglie proposed that every matter has accompanying wave (matter wave) [14]. Matter wave was confirmed experimentally by measuring diffraction pattern of electron, awarding a Nobel Prize to de Broglie in 1929. Matter wave could be even measured from macromolecules such as fullerene with C_{60}.

De Broglie further suggested that the wave inherent to matter is constantly moving through the matter guiding the trajectory of particle, and called it as pilot wave [15]. However, pilot wave has been long time forgotten until rediscovered by Bohm in 1952. Bohm developed pilot wave theory into what is now called the de Broglie-Bohm theory [16]. The de Broglie-Bohm theory suggests that underlying the observed probabilistic nature of the universe (general view of quantum science) is a deterministic objective property, which exists as hidden variable. The de Broglie-Bohm theory is now considered to be a valid challenge to the prevailing orthodoxy of the Copenhagen interpretation, but it remains controversial.

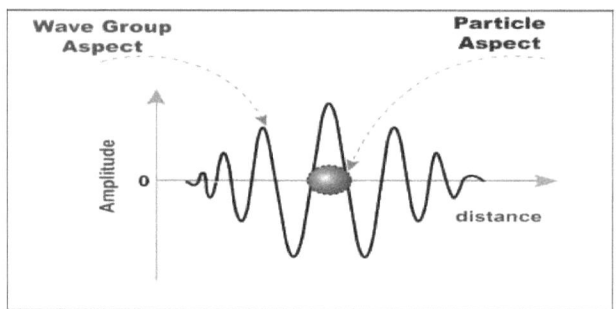

Pilot wave constantly move through from the rear to the front

10. Pilot wave faster than the speed of light forms 3D wave

In 1961 Eisberg showed by calculation that pilot wave is much faster than the speed of light [17]. Pilot wave of mass particle moving below the speed of light should be faster than the speed of light by following equation:

$$w = \frac{c^2}{v_p}$$ w: speed of pilot wave, v_p: speed of particle, c: speed of light

For a positive mass particle which is travelling at $v_p < c$, we see a set a waves moving faster than the speed of light. According to Tiller, pilot waves are always constantly moving through the matter form the rear to the front, just as occurs in water waves, directing a mass particle [18]. Calling such wave as 'information wave', as it does not spread out like electromagnetic wave and rather decrease thermodynamic entropy (negative entropy), Tiller suggested mass particle and information wave interacts so as to be experimentally operational. Pilot wave is expressed as 3D wave in this book, as it forms a field with 3D structure by constantly moving around the matter

11. Water memory explained by 3D wave

According to Tiller, pilot wave of the matter could be transferred to water by physical stimulation as in homeopathy [18]. If pilot wave separated from the matter to water still constantly moves faster than speed of light, it would not spread out and decrease thermodynamic entropy. If anything which is not a particle does not spread out and affect other particle, it should be carrying out its effect as a field. Bohm expressed this field characteristic of pilot wave as 'quantum potential' which operates as hidden variable [16].

The structure of the field created by pilot wave in water must not be random. It would be reasonable to think that the field is similar to or at least related to that of original shape of the matter (expressed as 3D wave in this book). As previously shown in phantom effects, the field created by pilot wave is not confined to water.

The hypothesis regarding water memory could be summarized as following: The pilot wave of the molecule (3D wave) transferred to water could exist as separate entity with similar field structure to original molecule and interact with cellular receptor, and then intracellular signal transmission could be initiated, which explains water memory.

12. Imaginary matter

If anything is faster than the speed of light, it should have negative mass according to special relative theory which does not allow speed beyond the speed of light.

$$m = \frac{m_o}{\sqrt{1 - v_p^2 / c^2}}$$ where m is the total relativistic mass and v_p is the velocity of the particle.

According to Dirac, space is full of particles with negative energy and negative mass (Dirac sea of vacuum) [19]. This corresponds to the area of imaginary number. The world consists of real area with positive energy and imaginary area with negative energy. When enough energy is given to the particle with negative energy in imaginary area, it jumps to real physical area, leaving a hole in imaginary area. A hole in imaginary area corresponds to anti-matter. Only infinitesimal part of space was converted to matter.

Matter consists of real part (physical area) and imaginary part. Imaginary part of the matter (called as imaginary matter in this book) could not exist like the particle in physical world but carry out its effect by creating a field by constantly moving around the matter. When imaginary part was separated from original matter in water, it still could interact with other matter as demonstrated in water memory.

Combining both pilot wave theory and Dirac sea of vacuum, it could be suggested as following: Matter consists of both real part and corresponding imaginary part which is faster than speed of light (pilot wave or 3D wave in this book). Imaginary part of the matter could be separated from the matter and interact with other matter (or imaginary part of other matter).

16

If the interaction between imaginary matters is possible, it might be even possible to hypothesize that real part of matter provides a frame and the imaginary parts of the matter provides function.

13. Digital Biology

Benveniste further showed that water memory could be recorded to computer by passing white noise through the solution: an aqueous solution in which molecules were dissolved was put into a copper tub, then white noise was applied to one side of the wall of copper tub and it was recorded from the opposite side of the wall of copper tub using a microphone which can record sound waves of 20 to 20,000 Hz. Benveniste and colleagues confirmed through repeated experiments that, when the recorded sound wave was converted into a vibration signal to vibrate the water using transducer, a physiological reaction was also induced [20-23].

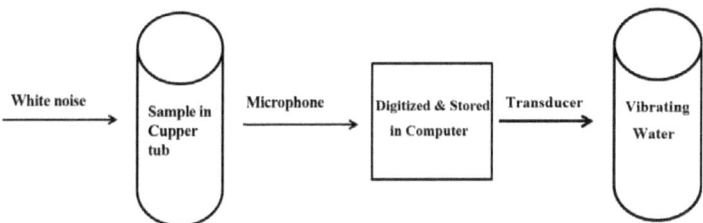

Digitization of 3D wave of the matter using white noise

Benveniste even showed that the recorded and digitized sound file could be transferred through email to induce physiological reaction in the far distance [24] As biological reaction could be regenerated even from the computer recorded digitized signal, he developed his theory under the name *Digital Biology* [25]. *Digital Biology* inducing physiological reaction without the presence of any material and digital recording/transmission of the substance wave could be reproduced in other laboratories [26-28].

14. Digitized 3D wave expressed in 2D space

3D wave of the substance cannot be detected by current scientific technique. However, 3D wave of the substance could be accessible to human senses when it is modulated to human familiar wave pattern such as sound or light. Benveniste used white noise to pick up the 3D wave of the substance. In this article visualization is preferred over the sound.

Homemade electronic device using 7.8Hz frequency and strong light source with visual range were used to digitize 3D wave of substance. The computer recorded digitized signal of the substance could be expressed as a visual shape in 2D space in a way of printing to plastic card or other flat surface, which should be much easier to use and have wider applications over the white noise [29].

Digitized 3D wave expressed in 2D space could also induce the same physiological reaction as that of original substance. For example, a plastic card containing digitized 3D wave of medically effective substance such as drug or hormone could be carried close to human body to induce physiological effect. It could be also possible to attach card on a water bottle to transfer 3D wave of the specific substance to water. Such digitized 3D wave could be expressed in any 2D space such as on clothes and on walls to affect environment [29]. Digitized 3D wave is explained in detail at the part III of this book.

Digitization gives great advantage over analogue type water memory, in that it stops deterioration, while 3D wave of the substance transferred to water gradually disappears depending on time.

15. 3D wave connected through holographic space

How could a printed shape in flat surface elicit the same biological reactions as the original substance? It may be possible that digitized 3D wave expressed on plastic card is connected with the original substance through holographic space to form 3D structure around the card.

Hologram refers to the 3D image created by the interference of two kinds of laser beams. The very special characteristic of hologram film is that it generates a whole shape when radiated with laser beam even if the film is cut into small pieces. The difference in shapes from whole film and cut film is just the resolution of image.

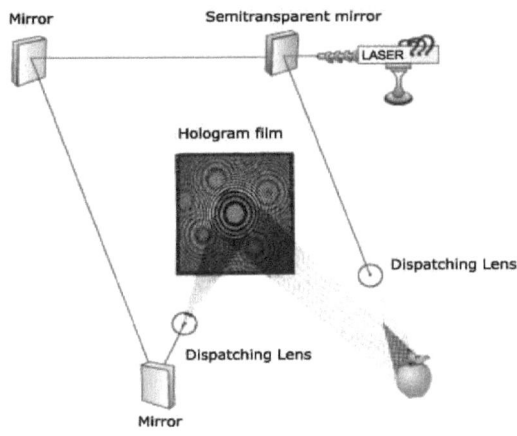

3D image by hologram

A piece of hologrmal contains information of the whole

Many physicists are suggesting a theory that whole universe is interconnected by holographic principle [30]. In holographic space all points are equal to all others. It is meaningless to separate some position to another. There is no locality in hologram (non-locality). Every particle in holographic space is interconnected in superluminal and super-spatial way creating non-locality.

It is possible that digitized 3D wave expressed in common plastic card is commanding to the space itself (infinite sea of particle with negative mass with superluminal velocity) to form the 3D wave of the original substance around the card through hologram.

16. 3D wave effective for brain diseases

3D wave of the substance expressed in 2D space or stored in water could be especially effective in for curing brain disease. Brain is protected by blood brain barrier (BBB) which inhibit drug from outside to pass through brain. For example, serotonin deficiency is related to depression. However, there is no way to put serotonin through brain because serotonin could not pass through BBB. Thus, it is needed to bypass BBB.

SSRI (Selective Serotonin Reuptake Inhibitor) class antidepressant drugs are representative bypass drugs and are most widely prescribed currently. They delay the reuptake of serotonin resulting in serotonin persisting longer in brain. Although SSRI class antidepressant drugs are regarded as relatively safe drugs, they have serious side effects such as to increase the risk of suicide and to increase tendency for violence [31]. However, if 3D wave of serotonin itself could be given to patient using digitized 2D card or using memorizing ability of water, we can imagine that it could be an ideal safe drug, as there is no substance causing unwanted side effects like SSRI class.

Other than serotonin virtually every neurotransmitter which cannot pass through BBB could be medically utilized with the digitization technology or using memorizing ability of water.

It is being observed that many patients with diseases related with brain such as depression, anxiety, insomnia, dementia, Parkinson's disease and even ADHD (Attention Deficit Hyperactivity Disorder) and autistic children get improved with digitized 2D card and/or drinking water with 3D wave of corresponding neurotransmitter [32, 33].

17. 3D wave modulated to electricity

Digitized 3D wave of the substance could be modulated even to electricity and could be used to change the characteristic of electricity [29]. When digitized 3D wave of the substance which is known to neutralize electromagnetic wave has been modulated into electricity, harmful nature of electromagnetic wave has been changed in beneficial way to the human and every electrical device could be used as a generator of specific 3D wave of medically effective substance.

Various effects could be expected from electrical devices by modulation of digitized 3D wave to electricity. Brain wave is stabilized while watching computer monitor. It is also noticed that the symptoms such as shoulder pain or eye congestion occurred by long time watching computer disappeared [29].

Electrical cooling system does not disturb balance of autonomic nervous system and thus does make people feeling not tired. It is also observed that food made by electrical cooker or microwave tastes better and freshness maintained longer. Especially, mobile phone gives people good feeling and does not get hot even after long usage. If mobile phone does not get heat, it means that battery can be used longer without losing energy as useless heat.

Characteristic of electricity modulated with 3D wave is explained in detail at the part III of this book.

18. 3D wave as a food contained in various dishes

We started from water memory and ended up with digitized 3D wave in 2D space creating 3D field around the space. Digitized 3D wave could be modulated to even electricity. 3D wave of the substance could be separated from original matter and could be modulated to various media. 3D wave of medical effective substance transferred to water, space, and electricity (including electromagnetic wave) can help us maintain health just by changing the environment.

It is like that food as 3D wave of the substance to be contained in a variety of dishes, which are water, space and electricity. The substance itself might be one of the containers for carrying the 3D wave of the substance (maybe the best container).

19. Characteristics of 3D wave summarized.

Characteristics of 3D wave can be summarized as the following:

1) Every matter has accompanying 3D wave.

2) 3D wave can exist without matter, but not in opposite way.

3) 3D wave of the matter can be transferred to water via physical stimulation

4) 3D wave of the matter in water interact with other matter.

5) Signal transduction of cell is via interaction between 3D waves.

6) 3D wave can be transferred to variety of other media than water.

7) The matter is the best medium for the corresponding 3D wave.

8) 3D wave of the matter can be digitized.

9) Digitized 3D wave expressed in 2D space shows same functionality.

10) Digitized 3D wave can be modulated to electricity

20. Another explanation: vortex theory

So far in this article I have assumed the existence of inherent wave part of the substance (3D wave) which could be separated by physical stimulation and exist independently in water and function like the substance.

In contrast Meyl suggested that every wave including electromagnetic wave could be rolled to vortex and then act like particle [34]. According to Meyl, everything is in wave state with only differences in the shape of the wave whether it is open or closed form (vortex), in which particle state is a highly condensed vortex, and thus, particle and wave state of single substance does not coexist, although their proportion is in equilibrium as a whole.

Vortex theory could also explain water memory, of which closed form of wave from the matter (vortex state) shows particle-like characteristic inside water [34]. However, vortex theory cannot explain the effect of digitized 3D wave expressed in 2D space. It might be possible that vortex itself could be carrier of 3D wave of the substance.

21. Characteristic of the wave depends on wavelength

The product of frequency and wavelength is the speed of light in general electromagnetic wave. Thus, frequency of electromagnetic wave depends on wavelength and vice versa. However, if the speed of 3D wave is not confined to the speed of the light, frequency and wavelength relation should be meaningless.

Meyl suggested that the characteristic of electromagnetic wave depends on wavelength regardless the speed of light [34]. Although 3D wave of the substance is faster than speed of light, it should have its own wavelength, regardless its speed. It might be possible to simulate the effect of the substance by radiating specific electromagnetic wave which has the same wavelength to that of 3D wave of the substance.

The following research shows that electromagnetic wave could be stored in water, and electromagnetic wave stored in water could be effective therapy for specific diseases.

22. Electromagnetic wave could be stored in water

British research team led by Dr. Smith published a very intriguing paper [35, 36]. They tested with patients having severe allergic reactions toward electromagnetic waves and also tested allergy-inducing substances diluting on homeopathic principles, and patient's reactions were observed.

Like the results of Benveniste, patient's reactions showed a phase of rise-fall repetition in the shape of a wave depending on the degree of dilution in the allergy-inducing substances.

They could even find frequencies in microwave area which induced severe allergies. The amount of allergy induction also resulted in the same changes showing the rise-fall repetition as frequency was changed. In addition, there found frequencies which eliminate allergic symptoms induced. The research team radiates the frequency onto water, and the patient's allergy symptoms also disappeared just by drinking the water. The ability of water eliminating allergy symptoms was maintained for at least six weeks.

Smith and his colleagues proved that water can remember the electromagnetic wave by showing patients can be treated by drinking that water.

23. Electromagnetic cellular communication

Kaznacheyev in the Soviet Union showed that any cellular disease or death pattern can be transmitted electromagnetically [37].

In the Kaznacheyev experiments, two sealed containers were placed side by side, with a transparent window separating them. Cultured tissue cells were separated into two identical samples, and each sample placed in each of the two halves of the apparatus.

The cells in one sample (on one side of the window) were then given small amount of harmful agents, leading to death of the exposed cell culture sample in 2-4 days.

If the thin window was made of ordinary window glass, the uninfected cells on the other side of the window were undamaged and remained healthy. However, if the thin optical window was made of quartz, sometime after the exposure (usually about 12 hours), same pattern of death appeared in the uninfected sample.

They found that cultured cells gave off photons in the near ultraviolet when they died. The normal window glass was opaque to these UV photons and absorbed them. In that case, the uninfected culture on the other side of the glass was not exposed to radiation by the UV (which carrying death signal) and they remained serenely healthy. However, when quartz window which was transparent to UV was used, death information carried by UV caused uninfected cells to die with same pattern. After repeating experiment more than 10,000 times, he could conclude that cellular death and diseases patterns can be transmitted and induced electromagnetically.

Cellular communication using UV light was earlier reported by Gurwitsch in 1930's. He found that cell divisions (mitosis) of onions were synchronized by the ultra-weak light passing through quartz but blocked by ordinary glass. He believed this light is in the range of UV and named the light leading to cellular communication as 'mitogenetic radiation'.

Mitogenetic radiation was later confirmed by the development of PMT (Photo Multiplier Tube) technology [38].

Using PMT (Photo Multiplier Tube) Popp could detect the light emitted from cells. He could demonstrate that cells communicate using the light in the range of 200-800 nm, which is a visible and near UV range. Popp named the light as bio-photon. As intensity of bio-photon causing cellular communication is very weak, he thought that bio-photon carry information rather than energy. According to Popp, every biological reaction is accompanied by bio-photon [39].

All these results suggest that very complex bio-information like the state leading to death or signal for mitosis (possibly complex mixture of many 3D waves) could be modulated to electromagnetic wave (living system uses UV for carrier).

24. Light in the beginning!

Matter could be converted to energy following the famous equation, $E=MC^2$, and vice versa. In the beginning it is believed particle was formed from energy. After first particle (elementary particle) is formed, more complex materials are followed, finally leading to life. Modern physics focus on the formation of the particles from energy. However, they do not ask why such relations takes place leading elementary particle to life in the direction of decreasing entropy.

There might be two kinds of lights in the beginning. The visible (detectable) light to form matter and is confined to the speed of light, which is the light we have known. The other light is the inherent wave of the matter (3D wave) described in this book which is faster than the speed of light and is always constantly moving through the matter form the rear to the front, directing the matter. 3D wave of the molecule can be separated from the matter and exist as separate entity and interact with other matters. Space itself is full of 3D waves leading to relations.

The matter is always accompanied by 3D wave. However, 3D wave can exist as separate entity separated from the matter. We need to ask which one is the first light in the beginning.

25. Era for new paradigm beyond materialism

As was shown in this article, there are countless facts which cannot be explained with conventional paradigm. What is needed is the scientific breakthrough like Plank did 100 years ago to explain black body radiation.

Current paradigm views world as materialism. Every biological reaction occurs by physical contact between molecules. If we scale the size of the cell as big city like Seoul where more than 10 million people live, the size of each molecule almost corresponds to each person. The meeting between two people without prior appointment when and where would be almost impossible. The biological reaction by physical contact between molecules cannot explain the fast reaction of living body.

If the interaction between imaginary matters (3D wave) could lead to cellular signal transmission, it does not need physical contact between molecules. If the interaction between matter and imaginary matter, or between imaginary matters is fast and good enough to induce cellular signal transduction, there is no need to rely on the materialism which cannot explain fast biological reaction.

Now we have seen that biological reactions could occur without presence of materials. There are enough facts which require new paradigm out of materialism. 21st century should be an era for new paradigm beyond materialism.

26. References

[1] http://en.wikipedia.org/wiki/Gödel's_incompleteness_theorems

[2] Clausius, N., Linde, K., Ramirez, G., Melchart, D., Eitel, F., Hedges, V.L. Are the clinical effects of homeopathy placebo effect? Meta-analysis of placebo controlled trials. **1997**, *The Lancet*, 350, 834-843.

[3] Dayenase, E., Beauvais, F., Amara, J., Oberbaum, M., Robinzon, B., Miadonna, A., Tedeshit, B., Pomeranz, P., Fortnerg, P., Belon, J., Saint-Laudy, B., Poitevin, B., Benveniste, J. Human basophil degranulation triggered by very dilute antiserum against IgE, *Nature*, **1988**, 333, 816-818.

[4] Schiff, M. The memory of water: Homeopath yand the battle of ideas in the new science. Thorsons, **1995**.

[5] Hardi, L., Arnoux, B., Benvenist, J. Effect of dilute histamine on coronary flow of isolated guinea pig heart, *FASEB,* **1991**, 5, A1583.

[6] Benveniste, J., Davenase, E., Ducot, B., Spira, A. Basophil achromasia by dilute ligand, *FASEB*, **1991**, 5, A1008.

[7] Beneniste, J., Arnoux, B., Hardi. L. Highly dilute antigen increase coronary flow of isolated heart from immunized guinea pigs, *FASEB*, **1992**, 5, A1610.

[8] Belon, P., Cumps, J., Ennis, M., Mannaioni, P.F., Roberfroid, M., Sainte-Laudy, J., Wiegan, F.A.C. Histamine dilutions modulate basophil activation, *Inflammation Research,* **2004**, 53, 5, 181-188.

[9] Milgrom, L. "Thanks for the memory" *The Guardian* (3.15, 2001)

[10] Ennis, M. Basophil models of homeopathy: a skeptical view. *Homeopathy,* **2010**, 99, 1, 51–56.

[11] Kim, H. W. New approach controlling cancer: water memory. *Journal of Vortex Science and Technolgy*, **2013**, 1, 1-8.

[12] Montagnier, L, Aissa, J, Del Giudice, E., Lavallee, C., Tedshi, A., Vitiello, G. DAN wave and water. **2011,** *Journal of Phys*ics / Conf. Ser. 306 **(**http://arxiv.org/pdf/1012.5166)

[13] http://www.papimi.gr/poponin.htm.

[14] http://en.wikipedia.org/wiki/Matter_wave

[15] http://en.wikipedia.org/wiki/Pilot_wave

[16] http://en.wikipedia.org/wiki/De_Broglie-Bohm theory

[17] Eisberg, R. M. Fundamentals of Modern Physics, John Wiley and Sons, Inc.,New York, **1961**, 140-146

[18] Tiller, W. Science and Human Transformation, *PAVIOR*, **1997**, www.tiller.org.

[19] http://en.wikipedia.org/wiki/Paul_Dirac

[20] Aissa, J., Littime, M.H., Attias, E., Benvenste, J. **1993**, Molecular signaling at high dilution or by means of electronic circuitry, *Journal of Immunology*, 150, 146A.

[21] Benveniste, J., Aissa, J., Litime, I., Tsangaris, G., Thomas, Y. **1994,** Transfer of the molecular signal by electronic amplification, *FASEB*, 8. A398.

[22] Benveniste, J., Jurgens, P., Aissia. J. **1997**, Digital recording/transmission of the cholinergic signal, *FASEB*, 10. A1479.

[23] Thomas, Y., Schiff, M., Belkadi, L., Jurgens, P., Kahhak, L., Benveniste, J. **2000**, Activation of human neutrophils by electronically transmitted phorbol-myristate acetate, *Medical Hypotheses* 54, 1, 33-39.

[24] Benveniste, J., Aissa, J., Jurgens, P., Hseuh, W. **1997**, Transatlantic transfer of digitized antigen signal by telephone link, *Journal of Allergy and Clinical Immunology*, 99, S175.

[25] Benveniste, J. **1998**, *Digital Biology*: Specificity of the digitized molecular signal. *FASEB*, 10. A1497.

[26] Senekowitsch, F., Endler, P.C., Pongratz, W., Smith, C.W. **1995**, Hormone effects by CD record/ replay, *FASEB*, 5. A2270.

[27] Citro, M., Endler, P.C., Pongratz, W., Vinattieri, C. Smith, C.W., Schulte, J. **1995**, Hormonal effects by electronic transmission. *FASEB*, 5. A2271.

[28] Endler, P.C., Pongraz, W., Wijk, R.V., Waltl, K., Hilgers, H., Brandmaier, R. **1994**, Transmission of hormone information by non-linear means. *FASEB*, 8. A2313.

[29] Kim, H. W. 3D wave expressed in two dimension, *Journal of Vortex Science and Technology*. **2014**, *in press*.

[30] http://en.wikipedia.org/wiki/Holographic_principle

[31] http://en.wikipedia.org/wiki/Serotonin

[32] http://cafe.daum.net/khwsupport

[33] Kim, H.W. Water Bluebird, *Bookscom*, **2009**

[34] Meyl, K. Scalar Waves, *Journal of Scientific Exploration* **2001**, 15, 199-205.

[35] Choy, R. Y., Monro, J. A., Smith, C.W. Electrical sensitivities in allergy patients. *Clinical Ecology*. **1987**, 4, 3, 93-102.

[36] Choy, R. Y., Monro, J. A., Smith, C.W. The diagmosis and therapy of electrical hypersensitivity, *Clinical Ecology*. **1989**, 6, 119-128

[37] Kaznacheyev, V. P., Distant intercellular interactions in a system of two tissue cultures. *Psychoenergetic Systems,* **1976**, 1, 141-143.

[38] Gurwitsch, A. A. A historical review of the problem of mitogenetic radiation, *Experientia*, **1988**, 44, 545-550.

[39] Popp, F. A. Biophoton emission, *Experientia*, **1988**, 44, 543-544.

Part II.

New Approach Controlling Cancer: Water Memory

It was revealed that every matter has its accompanying wave. The wave part of the matter (3D wave) contains information and functions like matter. The wave of the matter can be transferred to water physically by shaking or tapping (succussion), and thus serially diluted water have been used to stimulate natural healing power in traditional homeopathy. This way of transferring the wave part of the matter to water has been scientifically demonstrated by Benveniste and other researchers. In this study, instead of traditional homeopathic method a new electronic machine was devised to transfer the wave of matter to variety of medium including water.

P53 functions as a potent tumor suppressor. However, there is virtually no practical way to utilize the function of P53 clinically. If the wave portion (3D wave) of P53 could be transferred to water or any medium contacting water, various strategies could be possible. In this study, 3D wave of P53 was first transferred to UM (mixture of ceramic balls which makes alkaline reduced water), and then UM imprinted with 3D wave of P53 produce alkaline reduced water containing 3D wave of P53 by contacting water. The water containing 3D wave of P53 strongly inhibited cancer proliferation, showed anti-metastasis, and increased apoptosis. Water memory effect could be very useful for future cancer therapy.

1. Introduction

1.1 Water memory proved in scientific ways

In 1988 Benveniste and colleagues published a controversial article showing a biological reaction of ultra-high diluted solution, which could be called as 'water memory'. In the paper it was demonstrated that human basophil degranulation was triggered by extremely diluted antiserum against IgE [1]. Since then, they published many papers proving water memory effect under various experimental conditions [2-4]. As biological reaction in the absence of any effective molecules cannot be explained by conventional theory, the results of Benveniste and colleagues sparked many investigations of various seriousness. The most serious one was the research performed double blind by 4 independent European laboratories [5]. They thoroughly investigated the possibility of water memory using basophil activation by extremely diluted histamine, and they all obtained the same results supporting Benveniste.

1.2 Digitization of 3D wave of the matter using white noise

Benveniste first used homeopathic method to activate water by shaking with each dilution. Later he has developed a new technology as follows: An aqueous solution in which molecules were dissolved was put into a copper tub, then white noise was applied to one side of the wall of copper tub and it was recorded from the opposite side of the wall of copper tub using a microphone which can record sound waves of 20 to 20,000 Hz. Thereafter, Benveniste and colleagues confirmed through repeated experiments that, when the recorded sound wave was converted into a vibration signal to vibrate the water using a transducer, a physiological reaction was also induced [6-10].

Benveniste further showed that recorded sound wave file could be sent through email and transferred sound signal could also induce physiological reaction by physically vibrating water using transducer [11].

These results suggest that there is inherent wave-like characteristic for each molecule, which could be transferred to water and could reproduce physiological reaction like the molecule itself. The inherent wave-like characteristic which interacts with environment was called as pilot wave by de Broglie and Bohm [12] and as information wave by Tiller [13], as it decreases thermodynamic entropy. As the structure of the field created by pilot wave in water should be similar to or at least related to that of original shape of the matter, it is expressed as 3D wave in this book.

1.3 3D wave transferring machine

In this research a new electrical device was devised to replace time-consuming homeopathy which needs repeated dilution with physical stimulation at each dilution [14].

The device uses 7.8 Hz frequency as carrier of 3D wave which is the resonance frequency of earth. Subtle magnetic field generated by 7.8 Hz frequency could activate and transfer the 3D wave of the matter. Using the device 3D wave of hormones and other cytokines could be transferred to water and other medium like ceramic balls. In this paper it is shown that 3D wave of the hormone or cytokine transferred to ceramic balls could be passed to water indirectly by contacting water. Such water containing 3D wave of hormone functions like hormone in biological system [14].

1.4 UM produce mineral reduced water (MRW)

Water consists 70% of Human body. Water reaches every tissue of human body within 30 minutes after drinking. It even flows through blood brain barrier and has almost no side effect. If water itself could work as a radical scavenger, it would be an ideal antioxidant.

Instead of using electrolyzed alkaline reduced water a mineral combination (UM, whose pronunciation meaning healing mineral in Korean) was developed to produce alkaline reduced water with high pH (9.8) and low ORP (-250mv) (MRW, Mineral Reduced Water).

Compared to electrolyzed alkaline reduced water, producing MRW by UM is convenient and less expensive. Reaction of magnesium with water was slow and stable, and thus ideal for producing alkaline reduced water. For example, magnesium reacts with water as follow: $Mg + H_2O \rightarrow Mg^{2+} + 2e^-: 2e^- + 2H^+ \rightarrow 2H \cdot \rightarrow H_2$. A relative decrease of H^+ by generation of H_2 increases the concentration of OH^- in aqueous solution, which induces high pH and low ORP.

Component	Mg	Coral	Tourmaline	Illite	Biotite	Magnetite
Wt %	50	10	10	10	10	10

Composition of UM

1.5 Anti-oxidant effect of MRW

Alkaline reduced with high pH and negative oxidation reduction potential (ORP) was shown to have super oxide dismutase (SOD)-like activity and catalase-like activity, and thus, scavenge reactive oxygen species (ROS) which render oxidative damages to biological macromolecules and protect DNA from damage by oxygen radicals *in vitro* [15]. As ROS cause or aggravate variety of incurable diseases such as cancer, cardiovascular diseases, neuro-degenerative diseases as well as aging, if alkaline reduced water could work as a radical scavenger, it is believed to prevent and/or cure variety of diseases due to oxidative damage including cancer.

Alkaline reduced water (MRW) was produced with UM. Anti-oxidant effect of MRW was tested both *in vitro* and *in vivo* model.

MRW showed a concentration dependent anti-oxidant effect as the reduction of ROS by 50 % MRW solution (*vs* buffer) was similar to that of 10 µM ascorbic acid, well known anti-oxidant.

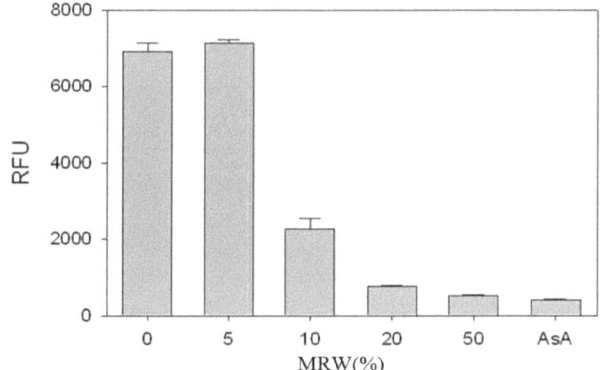

Antioxidant effect of MRW *in vitro*

Consistent with the *in vitro* data, the amount of ROS for lung, liver, and kidney were very low in MRW fed mice (C57BL/6) compared to that of control. However, the spleen, which is a major organ for immunity, shows higher levels of ROS in MRW fed group. This could be due to the immune boosting effect of MRW.

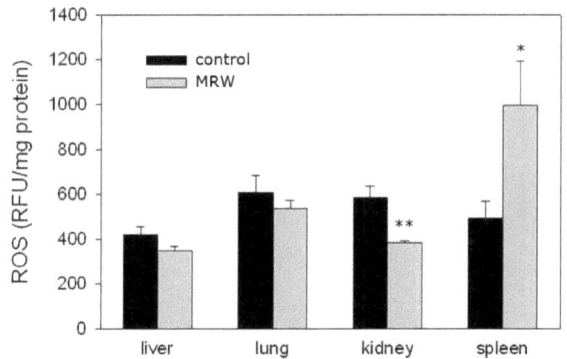

Antioxidant effect of MRW *in vivo*

1.6 Immune-boosting effect of MRW

Amount of ROS was significantly reduced in most of the organs in mice fed with MRW except for spleen. As spleen is a major organ for immunity [16], it is expected that MRW could boost immunity.

MRW intake showed time dependent rise of serum concentration of systemic cytokines, such as, Th1 (IFN-γ, IL-12, IFN: interferon, IL interleukin), cytokines for cellular immunity and Th2 (IL-4, IL-5), cytokines for humoral immunity, suggesting strong immune-boosting effect. Both Th1 and Th2 reached maximal peak after 2 week of MRW feeding and decreased back, but to the level still hi$\text{\textsf{ç}}$

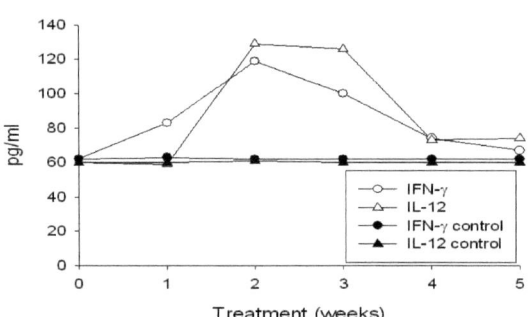

Effect of MRW on time kinetics of Th1

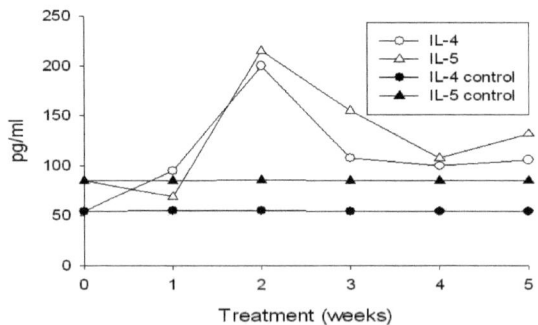

Effect of MRW on time kinetics of Th2

1.7 Rebounding effect explained by transient rise of immunity

Transient maximal peak of immunity after 2 weeks feeding of MRW to mice and gradually returning to almost to the level of base line after further feeding of MRW might explain so called 'rebound effect' accompanying transiently aggravated symptoms which is frequently observed during natural healing process.

The data indicate that mechanism of MRW could be multi-dimensional including anti-oxidant effect and enhancement of host immune function [16].

1.8 Anti-cancer effect of MRW

Considering that MRW acts as a scavenger of reactive oxygen species (ROS) and free radicals that contributes to cancer progression and that MRW provides strong immune-boosting effect, the possibility was explored for anti-cancer properties [16]. When melanoma cells were inoculated subcutaneously and intra-peritoneally into mice (C57BL/6), mice fed with MRW showed a significant delay in tumor growth and lengthened their survival compared to control group.

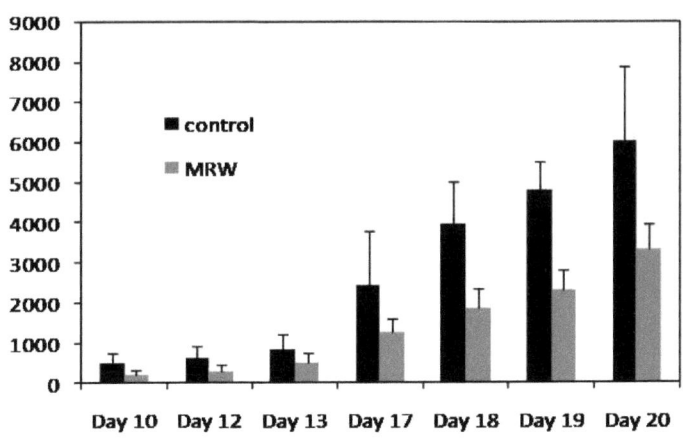

Effect of MRW on the inhibition of tumor proliferation

The inhibition of metastasis was also observed by MRW fed mice. The numbers of melanoma induced colonies in the lung were reduced when these cells were injected through tail vein.

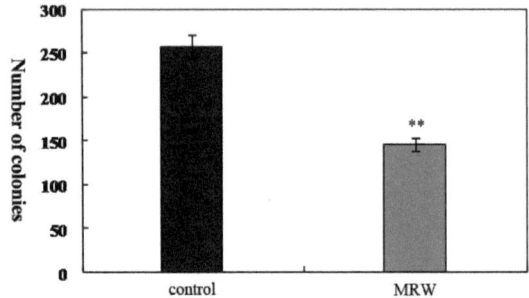

Effect of MRW on the metastasis of B16-BL6 melanoma cell on Lung

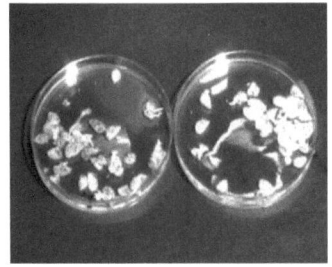

Picture showing reduced metastasized colonies by MRW fed mice (right)

1.10 Anti-diabetic effect of MRW

Anti-diabetic effect was demonstrated using OLTF (rats which had a diabetes-inducing hereditary defect as it grew) [17]. OLETF fed with MRW showed lower ARW glucose levels than control group fed with tap water. MRW also lowered blood triglyceride, and total cholesterol, suggesting that MRW would be effective for metabolic syndrome. Both hyperglycemia and hyperlipidemia are known to be related to the high ROS levels in blood vessels, tissues and cells [18, 19].

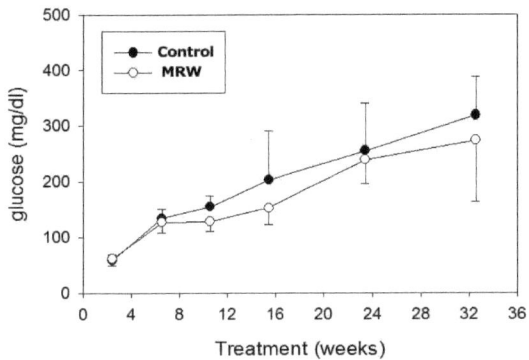

Effect of MRW on Glucose

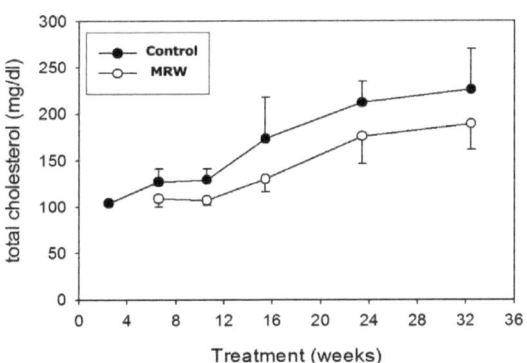

Effect of MRW on Choleterol

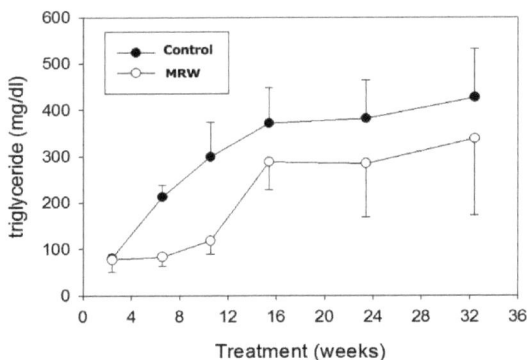

Effect of MRW on Triglyceride

1.11 MRW: Water Panacea.

There is no medicine which is effective both for cancer and diabetes, suggesting that the mechanism of MRW for is multi-dimensional, and thus different from that of current medicine effective for only specific disease condition. It is known that ROS and low immunity contributes to most of diseases. As MRW reduces ROS and increase immunity, it is expected that MRW should be effective for various diseases originated from high ROS and/or low immunity.

Water reaches every tissue of human body within 30 minutes after drinking. It even flows through blood brain barrier with no obstacle, and has almost no side effect. Taking MRW could be an ideal way to maintain health and also for therapeutic purpose.

1.12 3D wave of P53 transferred to MRW

P53, a transcription factor with molecular weight of 53 kilodalton, functions as a potent tumor suppressor [20]. P53 hold the cell at G1/S regulation point long enough for DNA repair proteins will have time to fix DNA damage and the cell will be allowed to continue cell cycle. P53 can also initiates apoptosis, the programmed cell death, if DNA damage proves to be irreparable. P53 is central to many of the cell's anti-cancer mechanisms. Thus, mutation in P53, could make it lose function as a tumor suppressor, which happens in most of the caner. Although P53 is so important for control of cancer, there is no way to utilize the function of P53 clinically. Most of the P53 researches are focused on gene therapy whose safety was not confirmed. However, if we could transfer the 3D wave of P53 to water, various strategies could be possible.

In this study we transfer the 3D wave of P53 to UM which makes alkaline reduced water. Alkaline reduced water produced by UM (MRW) with 3D wave of P53 was investigated for its anti-cancer effect.

2. Method

2.1 Generation of Mineral Reduced Water (MRW)

A special combination of ceramic balls plus magnesium (named as UM meaning healing mineral in Korean, patent no: KR10-074860 and KR10-068 1409) were devised to produce mineral alkaline reduced water (MRW). After overnight contact of 50 g of UM with 2 liter of tap water, MRW with pH 9.8 and -290 mV of ORP was prepared and used for the experiment.

3D wave of P53 was transferred to UM by homemade electronic device which uses 7.8 Hz frequency as a carrier. Subtle magnetic field was generated with the frequency around the input container where coil is wrapped around to activate substance. Activated 3D wave of P53 was transferred to UM which is in the output container where coil was wrapped around. 3D wave of P53 transferred to UM could be passed to water indirectly by contacting water.

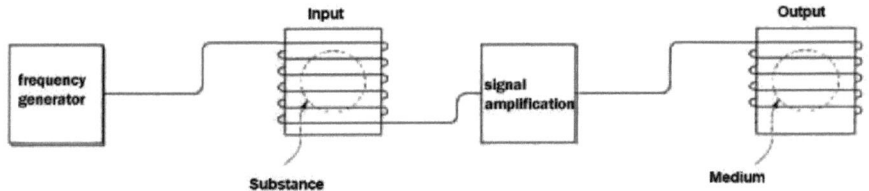

Block diagram of 3D wave transferring device

2.2 Cell culture

The MDA-MB-435 and MDA-MB-231 human cancer cells were obtained from the Lombardi Breast Cancer Depository at Georgetown University. They were grown in low glucose DMEM containing 10% FBS, 1 % penn/strep, and 25 mM HEPES. For MRW condition, UM was incubated with DMEM for 3 hours before addition of FBS, penn/strep. pH of culture media was readjusted to 7.8 to generate MRW-DMEM. All cell lines were cultured in humidified incubators at 37 °C in 5 % CO_2. For the cell proliferation assay, cells were plated on 6 well culture plate at the density of 2×10^5 cells per well and their growth was measured every other day by using bright line counting chamber (Hausser Scientific, Horsham, PA).

2.3 Apoptosis assay

The cells were seeded in 6-well culture dishes at a density of 1 to 3×10^5 cells per well in DMEM supplemented with 10 % FBS and were grown overnight at 37 °C in a humidified incubator with 5 % CO_2. Cells were treated with MRW or control media for 3 days, followed by apoptosis assay using the Annexin V-PE Apoptosis Detection Kit I (BD Biosciences, San Diego, CA).

2.4 Cell motility and invasion assay

For the cell motility assay, the upper and lower surface of the membrane in transwell inserts (Costar, Cambridge, MA) were coated with collagen I at 4°C overnight. To prepare for the invasion assay, matrigel (0.5   g, Collaborative Research, Bedford, MA) was diluted with cold water and dried onto each filter overnight at room temperature. On the following day, transwell membranes were blocked with DMEM for 1 hour at 37°C. Cells were trypsinized and resuspended in serum free DMEM/bovine serum albumin. A total of 10^5 cells were added to upper chamber of each well. 100 ng/ml lysophosphatidic acid (LPA) was added to the lower chamber as a chemo-attractant [23].

For MRW condition, cells were incubated with MRW-DMEM for 24 hours before the assay and during the assay. Inserts were incubated for 2-3 hours and non-migrating cells were mechanically removed using cotton swabs. The number of cells that were attached to the bottom side of the membrane were stained and counted using crystal violet. Assays were performed in triplicate and repeated several times.

3. Results

3.1 Effect on cancer cell proliferation

The data that MRW acts as an anti-oxidant suggest the possibility of its anti-cancer effect [14]. To address this issue we used MDA-MB-435 and MDA-MB-231 human cancer cell lines. These cell lines were well characterized for their malignant behaviors to induce tumorigenesis and metastasis. To assess the effect of MRW on cancer cell proliferation, MDA-MB-435 and MDA-MB-231 cancer cells were maintained in either regular or MRW media for 1-5 days and their rate of proliferation was monitored. Both MRW and MRW with 3D wave of P53 showed chemically same characteristic.

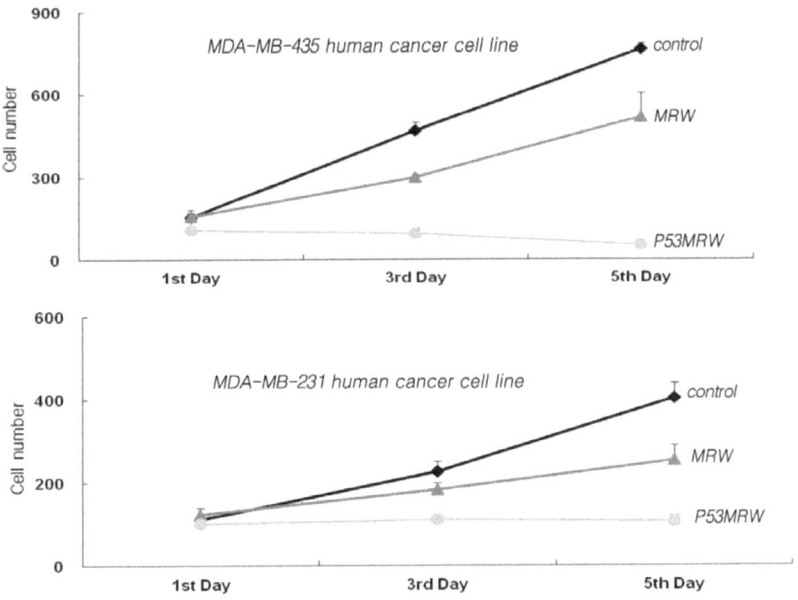

Effect of 3D wave of P53 on cancer cell proliferation

Control groups maintained a steady state growth as there is a 2 fold increase of proliferation in MDA-MB-435 cells and a 1.5 fold increase in MDA-MB-231 cells in every other day. In contrast, both MDA-MB-435 and MDA-MB-231 cells under MRW condition did show moderate declining in cell numbers. However, MRW with 3D wave of P53 showed marked decrease in cell number.

Data showed that MRW effectively blocks cancer cell growth, but when P53 information was transferred to MRW, cancer cell growth was almost completely blocked suggesting that the potent tumor suppressing effect of 3D wave of P53.

3.2 Effect on apoptosis

Declining of cancer cell growth by MRW suggests that the prolonged incubation of cancer cells with MRW may induce cancer cell death. To test the effect of MRW on apoptosis of cancer cells, we monitored the apoptotic index of MDA-MB-231 and 435 cells by measuring annexin PE staining under regular and MRW tissue culture conditions. Incubation of these cells with MRW for 3 days increases apoptosis of MDA-MB-435 cells about 5 fold and that of MDA-MB-231 cells about 2 fold.

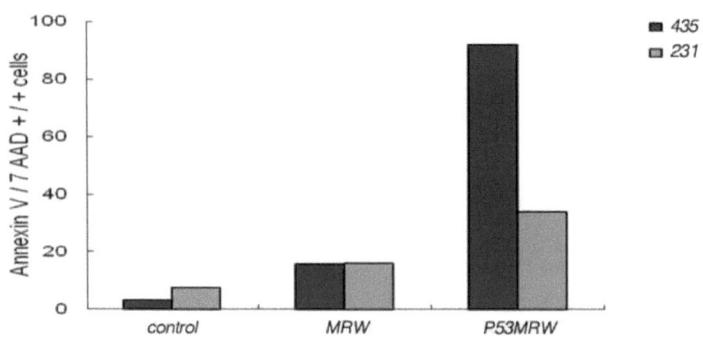

Effect of 3D wave of P53 on apoptosis

We did not observe a significant increase of apoptosis of these cell lines up until 2 days of MRW incubation (data not shown) suggesting that long term incubation (3 days or longer) of MRW induces apoptosis of cancer cells.

When incubation was carried out with MRW with 3D wave of P53, apoptosis markedly increased almost 30 fold for MDA-MB-435 cells about 10 fold for MDA-MB-231 cells, suggesting that information of P53 contained in UM worked effectively to increase apoptosis.

3.3 Effect on cancer cell motility and invasion

We next assessed the effect of MRW on cancer cell functions important for late stages of progression such as cell motility and invasion. Cell motility and invasion are the essential characteristics of cancer cells for metastasis [22].

We pre-incubated both MDA-MB-231 and MDA-MB-435 cells with MRW, MRW with 3D wave of P53, and control culture media for 24 hours prior to the cell motility and invasion assay. As mentioned previously, there was no significant increase in apoptosis during 24 hours of MRW incubation, which rule out the possibility that the effect on cell motility and invasion is due to cell death.

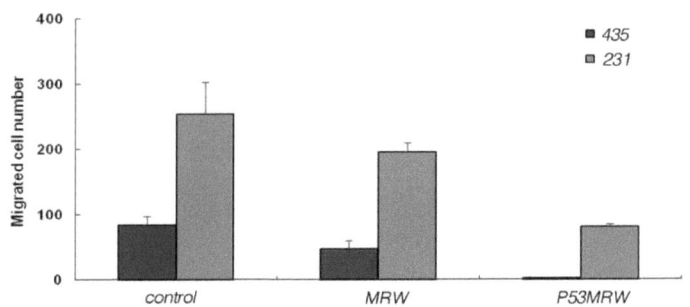

Effect of 3D wave of P53 on migration

MRW alone dramatically reduced the ability of MDA-MB-435 and

MDA-MB-231 cells to migrate towards LPA by about 75% and 90% respectively. Invasive potential of these cells was reduced to similar extent. Based on our findings, MRW may prove to be a potent anti-migratory agent that potentially prevents the spread of the breast cancer from primary origin to distal organs. Especially, MRW with 3D wave of P53 proved to be a much more potent anti-migratory agent. As both MRW and MRW with 3D wave of P53 have same chemical characteristics, it could be suggested that anti-tumor suppressing effect of 3D wave of P53 worked against the inhibition of migration

3.4 Effect on cancer cell progression in animal model

Luciferase-expressing cancer cells were constructed MDA-MB-231 and were injected into mice. Progression of cancer cells can be directly monitored using bioluminescent *in vivo* imaging. The picture shows MRW with 3D wave of P53 (right) strongly inhibited tumor growth

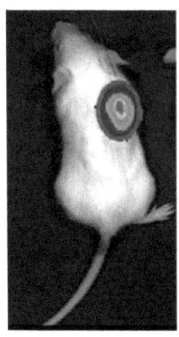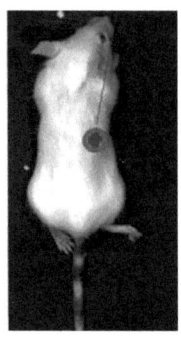

In vivo imaging of cancer cell

The following graph shows the time dependent proliferation of cancer cells (MDA-MB-231). Mice fed with MRW to which 3D wave of P53 was transferred shows significant delay in tumor growth compared to control group fed with normal tap water

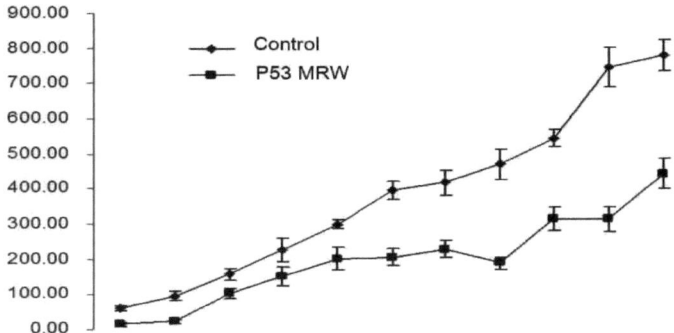

Effect of MRW with 3D wave of P53 on inhibition of tumor growth in mice.

4. Discussion

4.1 Homeopathy uses water memory

Homeopathy views diseases symptoms originated from the human body's natural healing process [24]. Homeopathy uses potential toxins which induce symptoms similar to diseases to invoke natural healing process. As the substances which induce disease-symptoms are mostly toxins with serious side effects, homeopathy used the dilution method until there are no harmful effects to the human body, but by physically stimulating (either tapping or shaking) with each dilution. Homeopathy claims that the effects did not disappear even after levels of dilution were done until none of the toxin's molecules were left in the solution. Homeopathy has been ignored by the orthodox medical circles until now because current science could not explain how a substance's effect can be shown without the actual substance. However, even though it was ignored by the orthodox medical circles, there have been almost three hundred verification experiments done on homeopathic effects in the past decade. Furthermore, about 80% of those showed homeopathy to have different effects from placebo effect. Even though not empathized, homeopathy uses water's memorizing ability [24, 25]. If homeopathy is effective therapy, it implies that water could store the 3D wave of toxins to boost natural healing power. If water could store 3D wave of material, then this capacity does not need to be confined to homeopathy.

4.2 Benveniste proved homeopathy in scientific ways

Benveniste was the first to show water memory effect in scientific way using human basophil degranulation [1]. According to Benveniste and colleagues, human basophil was still degranulated when anti-IgE was diluted extremely until there was no molecule left in the solution. Many debates were followawfterward, as water memory cannot be explained by conventional theory

48

in which molecules should come in physical contact with a cell receptor to initiate signal transmission.

In 2004 water memory effect was finally proved by double blind test using human basophil activation by extremely diluted histamine. This study was thoroughly investigated by 4 independent European laboratories [5]. They could not explain their findings and encourage others to investigate this phenomenon.

4.3 *Digital Biology*

Benveniste suggested that when a wave propagated from molecule is transferred to a cell receptor through water, the wave can induce resonance of a receptor initiating intracellular signal transmission. He even showed that the inherent wave of the matter could be recorded in digital form and could be transferred through email to induce physiological reaction in the far distance. Benveniste developed his theory under the name, *Digital Biology*.

3D wave of the matter could be modulated to variety of medium. Benveniste modulated the 3D wave of the matter to sound wave and the sound signal was recorded in a computer. Recorded digitized sound signal was converted into vibration signal using transducer to reproduce physiological reaction in the water [6-11].

4.4 Recent approaches on water memory

In this research a new electrical device was devised to replace time-consuming homeopathy [14]. The device uses 7.8 Hz frequency which is the resonance frequency of the earth to activate and transfer the 3D wave of the matter. Recently Montagnier and colleagues used similar device using 7.8 Hz frequency to transfer the information of DNA to water [26]. They showed that DNA polymerase could recognize DNA wave transferred to water and produce new DNA copies, suggesting that 3D wave of the DNA is the physical entity. The existence of the 3D wave of the DNA in space as physical entity was also shown in so called 'phantom DNA effect' by Poponin [27]. He showed that diffraction pattern of DNA which appear by laser radiation could be also generated even after DNA sample was removed.

4.5 3D wave of matter transferred to other medium than water

3D wave of the matter could be transferred to other medium like ceramic balls (UM in this book). The 3D wave of the matter could be modulated to even to electricity using the device. By applying alternating magnetic field generated by 7.8 Hz to electrical cord wound around a tub of a certain shape, the 3D wave of the matter in the input unit could be modulated to electricity [14]. Using the technology every electrical device could be turned into a generator of 3D wave for hormones or cytokines [28, 29].

4.6 3D wave explains water memory

The inherent wave of the particle is not a new concept. In 1924 de Broglie proposed that every matter has accompanying wave (matter wave). The existence of the wave inherent to matter was experimentally confirmed, awarding a Noble prize for de Broglie. De Broglie further suggested that the wave inherent to matter is guiding the trajectory of particle (pilot wave) [12]. In 1952 Bohm redeveloped almost forgotten pilot wave theory. According to Bohm, pilot wave of particle is affected by the environment and also affects the environment [12]. Pilot wave is expressed as 3D wave in this article

In 1961 Eisberg showed by calculation that pilot wave is faster than the speed of light [30]. Calling such wave as 'information wave', Tiller suggested mass particle and pilot wave interacts so as to be experimentally operational [13] According to Tiller, information wave of the matter could be transferred to water, and the information wave transferred to water could affect receptor by resonance, and intracellular signal transmission could be initiated [13].

4.7 Other explanation for water memory

In contrast, instead of inherent wave portion of matter, Meyl suggested that electromagnetic wave itself could be rolled to vortex and act like particle [31] According to Meyl, everything is in wave state with only differences in the shape of the wave whether it is open or closed form (vortex), in which particle state is a highly condensed vortex, and thus, particle and wave state of single substance does not coexist, although their proportion is in equilibrium as a whole. Vortex theory of Meyl could also explain water memory, of which closed form of wave from the matter (vortex state) shows particle-like characteristic inside water [31].

4.8 Water memory effective for various diseases

P53 is a transcription factor which interacts with specific position of DNA. Our data demonstrated that 3D wave of P53 contained in water also could function as a tumor suppressor. Although P53 is very important for suppression of tumor, so far there is no way to utilize the function of P53 protein clinically. The 3D wave of P53 contained in water could be very useful for cancer therapy, especially for brain tumor where blood brain barrier (BBB) blocks the flow of substances.

Water can reach to every organ of human body in 30 minutes without any obstacle. The 3D wave contained in water at least will not give unwanted side effects due to material decomposition. 3D wave of the matter that can be contained in water is not confined to P53 [14]. Virtually every kind of cytokines and hormones can be clinically utilized [28, 29]. Further researches regarding water memory and 3D wave of the matter are expected.

5. References

[1] Dayenase, E., Beauvais, F., Amara, J., Oberbaum, M., Robinzon, B., Miadonna, A., Tedeshit, B., Pomeranz, P., Fortnerg, P., Belon, J., Saint-Laudy, B., Poitevin, B., Benveniste, J. Human basophil degranulation triggered by very dilute antiserum against IgE, *Nature*, **1988**, 333, 816.

[2] Hardi, L., Arnoux, B., Benvenist, J, Effect of dilute histamine on coronary flow of isolated guinea pig heart, *FASEB,* **1991**, 5, A1583.

[3] Benveniste, J., Davenase, E., Ducot, B., Spira, A, Basophil achromasia by dilute ligand, *FASEB*, **1991**, 5, A1008.

[4] Beneniste, J., Arnoux, B., Hardi. L, Highly dilute antigen increase coronary flow of isolated heart from immunized guinea pigs, *FASEB*, **1992**, 5, A1610.

[5] Belon, P., Cumps, J., Ennis, M., Mannaioni, P.F., Roberfroid, M., Sainte-Laudy, J., Wiegan, F.A.C. Histamine dilutions modulate basophil activation, *Inflammation Research 53*, **2004**, 5, 181.

[6] Aissa, J., Jorgen, P., Litime, I., Behar, I & Benveniste, J. Molecular signaling at high dilution or by means of electronic circuitry, *Journal of Immunology* 150, **1993**, 146A.

[7] Benveniste, J., Aissa, J., Litime, I., Tsangaris, G., Thomas, Y. Transfer of the molecular signal by electronic amplification, *FASEB*, **1994**, 8, A398.

[8] Aissa, J., Littime, M.H., Attias, E., Benvenste, J. Electronic transmission of the cholinergic signal, *FASEB*, **1995**, 5. A683.

[9] Benveniste, J., Jurgens, P., Aissia. J. Digital recording/transmission of the cholinergic signal, *FASEB*, 10. **2000**, A1479.

[10] Thomas, Y., Schiff, M., Belkadi, L., Jurgens, P., Kahhak, L., Benveniste, J. Activation of human neutrophils by electronically transmitted phorbol-myristate acetate, *Medical hypotheses*, **2000**, 54(1) ,33.

[11] Benveniste, J., Aissa, J., Jurgens, P., Hseuh, W. Transatlantic transfer of digitized antigen signal by telephone link, *Journal of Allergy and Clinical Immunology*, **1997**, 99, S175.

[12] htp://en.wikipedia.org/wiki/Matter_wave,
http://en.wikipedia.org/wiki/Pilot_wave

[13] Tiller, W. Science and Human Transformation, *PAVIOR*, **1997**, www.tiller.org.

[14] Kim, H. W. Apparatus for delivering substance information, *Korean patent*, **2010**, 10- 2010-0012200, Kim, H. W. Characteristic and application of apparatus for transcribing 3D wave. *Journal of Applied Subtle Energy,* **2011**, 9, 2, 32.

[15] Shirahata, S., Kabayama, S., Nakano, M., Miura, T., Kusumoto, K., Gotoh, M., Hayashi, H., Otsubo, K., Morisawa, S., Katakura, Y. Electrolyzed-reduced water scavenges active oxygen species and protects DNA from oxidative damage. *Biochem Biophys Res Commun*, **1997**, 234 , 269.

[16] Lee, K. J., Park, S. K., Kim, J. W., Kim, G. W., Ryang, Y. S., Kim, G. H., Ch, H. W., Kim, S. K., Kim, H. W. Anticancer effect of alkaline reduced water. *J. intl, Soc. Life Info. Sci*. **2004**, 22, 2, 302.

[17] Jin D., Ryu S.H., Kim H.W., Yang E.J., Lim S.J., Ryang Y.S., Chung C.H., Park S.K., Lee K.J. Anti-diabetic effect of alkaline. reduced water on OLETF rats. *Biosci Biotechnol Biochem* **2006**, 70, 31.

[18] Strawn WB: Pathophysiological &clinical implications of AT(1) & AT(2) angiotensin II receptors in metabolic disorders: hypercholesterolaemia & diabetes. *Drugs* **2002**, 62, 31.

[19] Nishikawa T, Edelstein D, Du XL, Yamagishi SI, Matumura T, Kaneda Y, Yorek MA, Beebe D, Oates PJ, Hammes HP, Giradino I, Brownlee M: Normalizing mitochondrial superoxide production blocks three pathway

of hyperglycaemc damage. *Nature* **2000**, 404, 787.

[20] http://en.wikipedia.org/wiki/P53

[21] Landriscina, M., Maddalena, F., Laudiero G., Esposito, F. Adaptation to oxidative stress, chemoresistance, and cell survival. *Antioxid Redox Signal*, **2009**, 11, 2701. 2716.

[22] Price, JE., Polyzos, A., Zhang, RD., Daniels, LM. Tumorigenicity and metastasis of .human breast carcinoma cell lines in nude mice, *Cancer research,* **1990**, 50(3), 717. 21.

[23] Chung, J., Yoon, S., Lipscomb, E., Mercurio, AM. Met and the integrin can function independently to promote carcinoma invasion, *J. Biol. Chem.,* **2004**, 279, 32287- 32293.

[24] Reilly, D. The puzzle of homeopathy, *Journal of Alternative and complementary medicine*, **2001**, 7, 103-109.

[25] Clausius, N., Linde, K., Ramirez, G., Melchart, D., Eitel, F., Hedges, V.L. Are the clinical effects of homeopathy placebo effect? Meta-analysis of placebo- controlled trials, *The Lancet*, **1997**, 350, 834-843.

[26] Montagnier, L, Aissa, J, Del Giudice, E., Lavallee, C., Tedshi, A., Vitiello, G. DAN wave and water, **2011,** *J. Phys. / Conf. Ser.* 306 (http://arxiv.org/pdf/1012.5166)

[27] http://www.papimi.gr/poponin.htm.

[28] Kim, H.W. Life of Water: A Cure for Our Body, *Bookscom,* **2008**

[29] Kim, H. W. Digitized 3D wave expressed in two dimension. *Journal Vortex Science and Technology,* **2014**, *in press.*

[30] Eisberg, R.M. Fundamentals of Modern Physics, *John Wiley and Sons, Inc., New Yonrk, N.Y,* **1961**, 140-146.

[31] Meyl, K. Scalar Waves, *Journal of Scientific Exploration.* **2001**, 15, 199-205.

Part III

Digitized 3D wave Extend to Space and Electricity

Inherent wave of the substance (3D wave) could be transferred to water and other medium. The 3D wave separated from the substance could be digitized and expressed as two dimensional shapes. The research demonstrated that digitized 3D wave of the substance could maintain functionality of original substance. Digitized 3D wave expressed in 2D space (UN) could be used to deliver the 3D wave of the medically effective substance.

Digitized 3D wave could be modulated to electricity. It is known that electromagnetic wave is harmful to human. However, electromagnetic wave could be changed in human beneficial way with application of digitized 3D wave. Other than effects to human, electricity itself could be changed in effective wave. It was observed that mobile phone does not get very hot even after long usage and total electricity usage of a house to be reduced. This research also demonstrated that electricity could be used as a medium for delivering 3D wave of the specific substance. This suggests that every electrical device could be used as a generator of 3D wave of the medically effective substance. 3D wave of the substance modulated to water, plastic card, and electricity would have wide application.

1. Introduction

1.1 3D wave of toxin stored in water in homeopathy

Homeopathy uses toxic substance to enhance the ability of natural healing. Homeopathy view symptoms of a disease as part of the healing process, thus by adding the substance that causes these symptoms to healthy person, the patient's natural healing power increases and thus cures diseases. Substances causing diseases symptom are mostly poisonous so even though the original disease gets cured, the poisonous substance could cause more damage to the body.

Thus what homeopathy does is to dilute the poisonous substance until it is no longer harmful to the body and physically stimulate it at each dilution. Even though the poisonous substance gets diluted in several steps to the point where the poisonous substance is no longer exist, the natural healing ability of human body actually increases rather than decreases [1, 2].

1.2 Homeopathy is effective therapy, though not explainable

Homeopathy has been ignored by the orthodox medical circles until now because current science could not explain how a substance's effect can be shown without the actual substance.

However, even though it was ignored by the orthodox medical circles, there have been almost three hundred verification experiments done on homeopathic effects in the past decade. Furthermore, about 80% of those showed homeopathy to have different effects from placebo effect [1, 2].

Even though not empathized, homeopathy uses water's memorizing ability. Homeopathy showed that even if the substance does not exist inside water, the information of the substance can be recorded. If water could stores information of substance, then this capacity does not need to be confined to homeopathy.

1.3 Homeopathy proved in scientific ways

In 1988, Benvensite and colleagues published a controversial article showing a biological reaction of ultra-high diluted solution [3], which could be called as 'water memory'. In their paper it was demonstrated that human basophil degranulation was triggered by extremely diluted (10^{-120} of its original density) antiserum against IgE but only when the each diluting solution was shaken based on the homeopathic method. Benvensite's paper was the first to scientifically prove homeopathy which water can record substance's information despite the substance not being existent. Since then, they published many papers proving water memory effect under various experimental conditions [4-6].

As biological reaction in the absence of any effective molecules cannot be explained by conventional theory, the results of Benveniste and colleagues sparked many investigation of various seriousness. The most serious one was the research performed double blind by 4 independent European laboratories in 2004 [7]. They thoroughly investigated the possibility of water memory using basophil activation by extremely diluted histamine, and they all obtained the same results supporting Benveniste. The interesting fact was the purpose of their double blind test was to disprove Benvensite's research but instead proved that it was in fact true [8]. Even though it was already proven by double blind test, the scientific world still does not show interest in water's memorizing ability.

1.4 Digitization of 3D wave using white noise as carrier

Benveniste has further developed a new technology as follows: an aqueous solution in which molecules were dissolved was put into a copper tub, then white noise was applied to one side of the wall of copper tub and it was recorded from the opposite side of the wall of copper tub using a microphone which can record sound waves of 20 to 20,000 Hz. Thereafter, Benveniste and colleagues confirmed through repeated experiments that, when the recorded sound wave was converted into a vibration signal to vibrate the water using a transducer, a physiological reaction was also induced [9]. He further showed that recorded sound wave file could be sent through email and transferred sound signal could also induce physiological reaction by vibrating water [10-14].

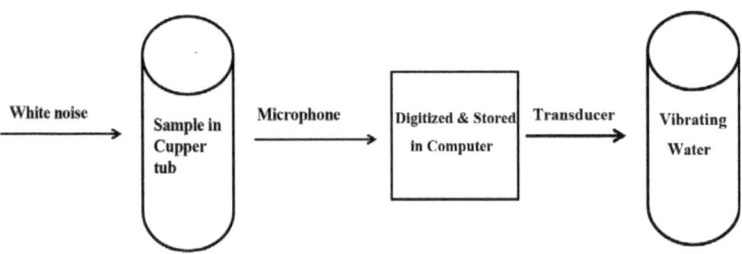

Digitization of 3D wave of the matter using white noise

As biological reaction could be regenerated even from the computer recorded digitized signal, he developed his theory under the name *Digital Biology* [15]. The previous theory was in order for signal transduction to occur, the hormone particles and the cell's receptor had to be physically meet. Benveniste suggested that when a wave propagated from molecule is transferred to a cell receptor through water, the wave can induce resonance of a receptor initiating intracellular signal transmission. Concepts of *Digital Biology* inducing physiological reaction without the presence of any material and digital recording/transmission of the substance' wave could be reproduced in other laboratories [16-18].

1.5 Pilot wave and 3D wave

These results suggest that there is inherent wave. like characteristic for each molecule, which could be transferred to water and could reproduce physiological reaction like the molecule itself. The inherent wave of the particle is not a new concept.

In 1924 de Broglie proposed that every matter has accompanying wave. The existence of the wave inherent to matter was experimentally confirmed, awarding a Noble prize for de Broglie [19]. De Broglie further suggested that the wave inherent to matter is guiding the trajectory of particle (pilot wave) [20].

In 1952 Bohm redeveloped almost forgotten pilot wave theory. According to Bohm, pilot wave of particle is affected by the environment and also affects the environment, which is whole universe [19].

In 1961 Eisberg showed by calculation that pilot wave is faster than the speed of light [21, 22]. Calling such wave as 'information wave' as it decrease thermodynamic entropy (also expressed as '3D wave' in this article), Tiller suggested mass particle and pilot wave interacts so as to be experimentally operational [22]. According to Tiller, pilot wave of the matter could be transferred to water by physical stimulation, and the pilot wave transferred to water could affect receptor by resonance, and intracellular signal transmission could be initiated [22].

1.6 Electrical method to transfer the 3D wave of the substance

We have also devised a new electrical device to transfer 3D wave of substance [23, 24]. The device uses 7.8 Hz frequency which is the resonance frequency of the earth to activate and transfer the wave part of the matter. Using the device 3D wave of hormones and other cytokines could be transferred to water. Such water containing 3D wave of the hormone functions like hormone in biological system [24].

For example, P53 protein functions as a potent tumor suppressor. However, there is virtually no practical way to utilize the function of P53 clinically. If the 3D wave of P53 could be transferred to water or any medium contacting water, various strategies could be possible. The water to which 3D wave of P53 was transferred strongly inhibited cancer proliferation, showed anti-metastasis, and increased apoptosis [25].

Noble laureate of 2008 Montagnier and colleagues recently showed that DNA polymerase could recognize DNA wave transferred to water and produce new DNA copies, suggesting that 3D wave of the DNA is the physical entity. They used similar device using 7.8 Hz frequency to transfer the information of DNA to water [26].

1.7 3D wave digitized and expressed into 2D space (UN)

Benveniste used white noise to carry the 3D wave of the substance. The white noise carrying 3D wave of the substance could be converted into digital form by recording into computer. The recorded signal could be easily converted into vibration signals using transducer to regenerate biological reaction in water.

However, this whole procedure of inducing biological reaction using white noise is inconvenient and difficult to apply.

In this research we have used 3D wave transferring device and strong light source to digitize 3D wave of the substance. Computer recorded 3D wave was expressed into a visual shape in two dimensional (2D) space. Digitized 3D wave expressed in 2D space showed the same functionality as the original substance, which is much easier to use and has wider applications.

2. Method

2.1 Electrical device transferring 3D wave of the substance

A new electrical device was devised to replace time. consuming homeopathy which needs repeated dilution with physical stimulation at each dilution (Figure 1) [23, 24] The device uses 7.8 Hz frequency which is the resonance frequency of earth. Using the device 3D wave of hormones and other cytokines could be transferred to water. Subtle magnetic field was generated with the frequency around the input container where coil is wrapped around to activate and transfer 3D wave of the substance.

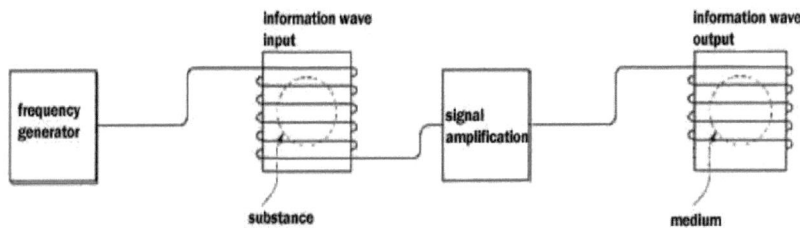

Block diagram of 3D wave transferring device

2.2 Digitization of 3D wave

We used fine iron powder to transfer the 3D wave of the substance. First, 3D wave of the substance is transferred to iron powder suspended with ethanol using information transferring device. Then iron powder suspension in ethanol are put in transparent crystal cuvette and exposed to strong light in visible range. The light passed through the cuvette is recorded and digitized using image sensor. Digitized images could be modified in computer.

2.3 Measurement of rotating electromagnetic wave

Rotating magnetic wave is measured at the Center for Rotating

Electromagnetic Wave of Ajou University at Suwon, Korea. The term 'rotating electromagnetic wave' means the waves released from a substance which should be another name for 3D wave. By measuring the magnetic field difference induced to an electromagnet by a rotating electromagnetic wave of the substance, the directional nature of the rotating electromagnetic (clockwise or counter clockwise) wave could be determined. The polar nature (positivity or the negativity) could be determined by exposing the substance to strong light and measure the amount of light adjusted by the substance [27, 28]. The positivity of a rotating electromagnetic wave indicates that the rotating radius decrease toward the direction of progress and the negativity indicates that the rotating radius increase.

2.4 Measurement of bio-resonance

Bio-resonance was measured using Quantum Resonance Spectrometer (QRS) at Conmaul Hospital of Oriental Medicine at Seoul, Korea. QRS is measuring interaction of 3D wave of the substance with human body which appears as a change of autonomic nervous system giving a numerical value for each code [29, 30]. Measurement with higher digit indicates more significance for each code. Bio-resonance measurement could be on the line of Radionics and could be expressed as more objectified version of the O-ring test or the muscle test used by Kinesiology measuring human interaction with substance [30]. It was reported that the confidence degree of bio-resonance measurement is over 95% for skilled operator [29].

2.5 Other measurements

^{17}O NMR was measured with 500MHz Avance. 500 (Bruker) at National Center for Inter. University Research facilities. Brain waves were measured with Neruoharmony system (Braintech).

3. Results

3.1 Rotating magnetic wave of UM and UN

There is currently no method to directly measure the 3D wave from substance. Recently it has been reported by that by characterizing the rotating electromagnetic wave it would be possible to predict possible effects of the substance to human body [27, 28]. The term 'rotating electromagnetic wave' means waves released from a substance which should be another name for 3D wave of the substance.

The 3D wave that is discussed in this book has been called as torsion wave, spin wave, and scalar wave, etc. in other places. Especially in Asia, these inherent 3D waves of the substance have been described as 'chi' for thousand years.

The following figure shows the measurement of rotating magnetic wave by UM (meaning healing mineral in Korean pronunciation). UM is a mineral combination to produce alkaline reduced water with high pH and low ORP developed [25].

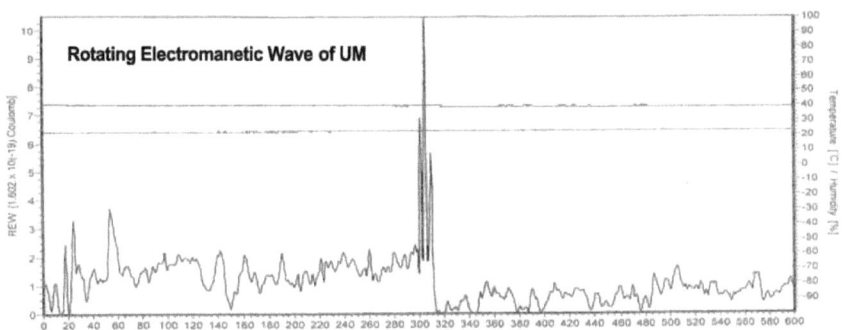

According to the pattern shown in the figure indicates that the rotating electromagnetic wave is being measured to counter clockwise rotational direction and its radius is being decreased towards the direction of progress (positivity). Out of 4 possible combinations counter clockwise & positivity pattern represents crystal bond which appears in natural rocks and is known to be human beneficial. The substance of this pattern is known to make human healthy, suppress tumor growth and harmful bacterial growth, and also neutralize the harmful nature of electromagnetic wave or water vein [27, 28].

The following figure shows the measurement of rotating magnetic wave by UN (meaning healing energy) card. UN is a digitized 3D wave expressed as a 2D shape in common plastic card.

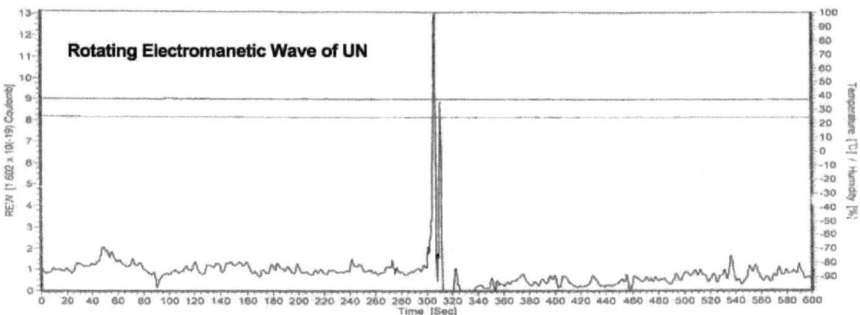

Surprisingly UN card containing the digitized 3D wave showed the same counter clockwise & positive pattern as the UM. General plastic card shows no pattern at all. This result shows that digitized 3D wave of the substance expressed in plastic card (UN) still maintains the same pattern of rotating magnetic wave as the original substance.

3.2 Bio-resonance measurement of UN card

3D wave of substance exists in the area of imaginary number which does not exist in the physical area, but carry out its effect as a field. Even though 3D wave cannot be measured directly with current scientific technique, its effect to human can be sensed. Bio-resonance is measuring interaction of 3D wave of the substance with human appearing as a change of autonomic nervous system giving a numerical value for each code [29, 30]. Bio-resonance measurement with higher digit indicates more significance.

Substance	UN card	Normal card
Serotonin	9	1
Dopamine	10	1

Measurement of neurotransmitter expressed in 2D card.

3D wave of medically effective substances such as hormones or neurotransmitter can be digitized and expressed in common plastic card (UN card). Bio-resonance of neurotransmitters expressed in plastic card showed high figures for its functionality compared to general plastic card used as a control. The results suggest that 3D wave of hormone or neurotransmitter digitized in 2D shape could have the same functionality as original substance.

3.3 Cluster size of water gets smaller by UN

When liquor goes through a long storage it tastes smoother. This does not mean that the alcohol level decrease but the cluster of alcohol and water is getting smaller, thus becomes easier to blend.

The decreased cluster of water and alcohol can be measured by the half linewidth of the water peak of ^{17}O NMR. Half linewidth of the water peak is known to be proportion to the size of the water cluster [30].

The following table shows the measurement of ^{17}O NMR of alcoholic liquor (20% v/v) before and after placing UN card under each bottle for an hour.

It is noticeable that the ^{17}O NMR's half linewidth has been decreased for alcoholic liquor by placing UN card under the bottle. These measurements suggest that due to digitized 3D wave expressed in UN card, cluster size of water and liquor gets smaller, thus easily mixes each other and makes the liquor taste smooth.

Sample	Half linewidth (Hz)
Liquor	183.9±1.9
Liquor (UN card)	161.7±4.6

Measurement of ^{17}O NMR half linewidth

3.4 Modulation of UN to electricity (UL)

3D wave of the substance could be modulated into electricity using 3D wave transferring device [24]. In this research the modulation of the 3D wave to electricity has been attempted by placing digitized UN (where 3D wave of the UM is expressed in plastic card) inside electrical cord wound around a tub where alternating magnetic field (220V, 60Hz) is formed.

After a mobile phone's battery has been charged with modulated electricity with digitized 3D wave (UL, meaning healing electricity), bio-resonance of male (age 46) has been measured while using mobile phone. Bio-resonance measurement with higher digit indicates more human beneficial.

Bio-resonance of a person using mobile phone that was charged with ordinary electricity showed negative figures of bio-resonance indicating very harmful to human, while that of mobile phone charged with UL showed rather higher numbers of bio-resonance even compared to that of not using mobile phone.

	Control (no mobile phone)	Mobile phone (ordinary battery)	Mobile phone (charged with UL)
Immunity	18	-17	22
Brain	18	-10	19

Bio-resonance of a person using mobile phone

These results suggest that it is possible to neutralize harmful effect of electromagnetic wave by modulating 3D wave of UM to electricity (UL). It might be possible that using mobile phones could rather improve human health.

Other than effects to human health, it is observed that mobile phone does not get very hot even after long usage. Although it was quite variable depending on the model of mobile phone, temperature measurement with i. phone4S after 30 minutes of continuous usage shows that temperature of mobile phone was 4.1 degrees lower when charged with UL, while the test with HTC mobile phone (EVO 4G+) showed 2.8 degrees difference. This could mean that electricity itself was changed in very effective way resulting in low resistance.

Temperature of mobile phones was measured before and after 30 minutes of continuous usage. Measurement was repeated 3 times first without application of UL, and then repeated 3 times after application of UL. Data are expressed as means±SD

(℃)	Mobile phone (ordinary battery)	Mobile phone (charged with UL)	Temperature difference
iphone4S	38.2±0.2	34.1±0.2	4.1
HTC EVO 4G+	37.4±0.6	35.6±0.1	2.8

Temperature difference of mobile phone with application of UL

Indeed total electricity usage of a house could be saved 23-38% with application of UL when compared to those of previous year. UL was installed to electricity distributing board of a house.

Total electricity usages of the houses after application of UL were compared with those of the same month of previous year. Data are expressed as KW per hour (KWH). UL was installed at the electricity distributing board of each house. Total electricity usage was reduced from 23% to 38% depending on the house with application of UL.

(KWH)	A	B	C	D
Before	184	128	325	445
After	261	208	232	343
%	70.4	61.5	71.4	76.5

Reduction of total electricity usage of the houses by UL.

The following figure shows time dependent change of electricity usage of a house after application of UL. Electricity usage of a house is almost same every year. UL was applied from June of 2013 for a house. Total electricity usage of a house could be saved almost 30% with application of UL

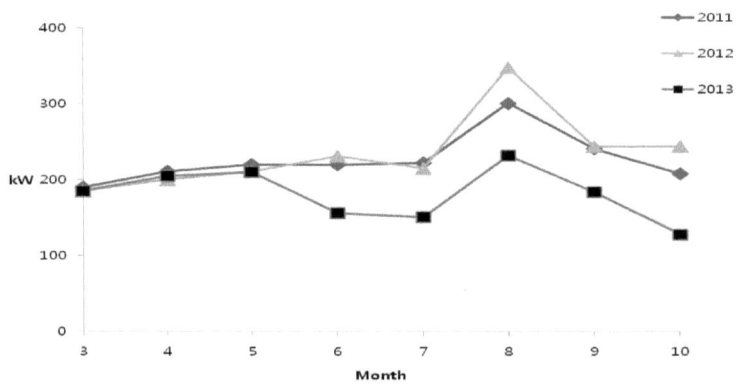

Time dependent change of electricity usage after application of UL

3.5 Modulation of specific 3D wave to electricity

Modulation of 3D wave of the substance to electricity could have wide application. It could be even possible to modulate the 3D waves of the specific substance into electricity. Serotonin is the largest single neurotransmitter system of the brain [31]. Its secretion and physiological actions mediate stress and pain, affecting both immune and nervous system functions. Serotonin dysfunction is well characterized in mental disturbances like depression and anxiety. We modulated the 3D wave of serotonin to electricity and test its effect by monitoring brain wave. While watching computer monitor connected to serotonin modulated electricity, brainwaves of various people were measured before and after application of UL.

While awake, delta waves and theta waves of brain should barely be appearing (the large peaks on the back side of the graph). To observe the influence of delta and theta waves, brain waves of a male (age 54, A) and female (age 54, B) were measured for 30 seconds, followed by 30 seconds resting period with eyes closed, and again measured for 30 seconds.

Male (age 54)

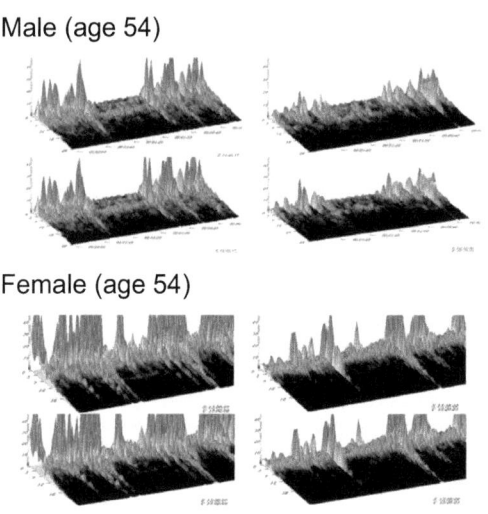

Female (age 54)

Measurement of brain wave watching computer with resting period

When eyes were closed, peaks of delta and theta wave have disappeared (for male) or reduced (for female). When delta waves and theta waves are active while awake, it means difficulty in focusing and easy distraction. It was observed that delta and theta waves of both male and female decreased with application of UL with 3D wave of serotonin.

In the following figure brain waves of various people were measured without resting interval with eyes closed. For every measurement the same pattern of brain waves were observed with application of UL. Delta and theta waves were active while staring at the computer monitor (left side of figure) but when staring at a monitor using electricity modulated with 3D wave of serotonin, intensity of the delta and theta waves decreased (right side of figure).

These results showed that 3D wave of the serotonin could be modulated to electricity and affect the brain wave of those who were staring at computer monitor. It is very likely that the modulation of the 3D wave of the substance should not be confined to serotonin. These results suggest that every electrical device could be turned into a generator of 3D wave for hormones or cytokines [25, 26] with application of UL.

A. Male (age 24)

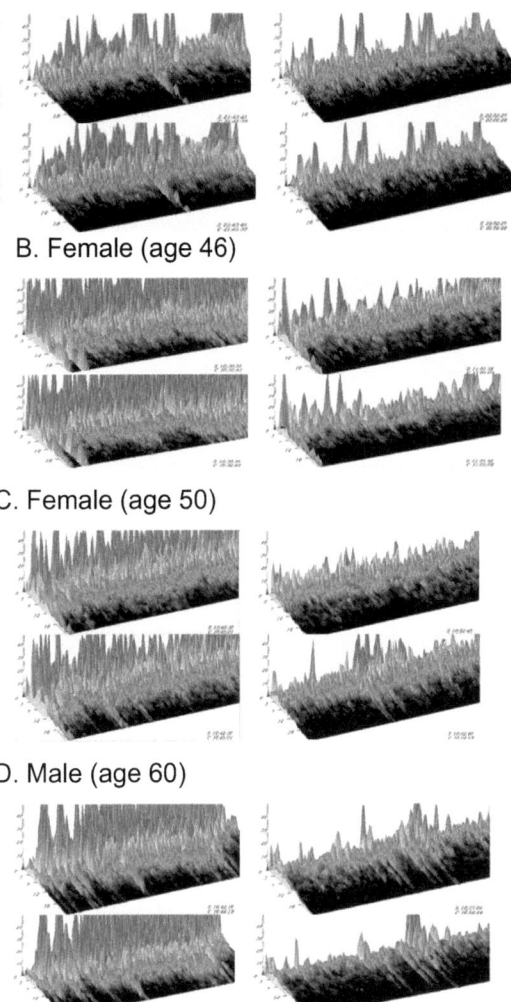

B. Female (age 46)

C. Female (age 50)

D. Male (age 60)

Measurement of brain wave while watching computer monitor

4. Discussion

4.1 3D wave of the substance expressed in 2D space

Benveniste modulated the 3D wave of the substance to sound and record the sound signal to a computer. Computer recorded digitized sound signal was converted into vibration signal using transducer to reproduce physiological reaction in the water [10-18]. In this book 3D wave of the substance was digitized and expressed as a certain shape in 2D space such as common plastic card (UN). Digitized 3D wave of the substance expressed in 2D space is much easier to use and has much wider applications. Although not proved through biological experiment, our results suggest that digitized 3D wave expressed in common plastic card could induce the functionality of the original substance.

Digitized 3D wave expressed in 2D space could be used in various ways. A card containing digitized 3D wave of medically effective substance such as drug or hormone could be carried close to human body to induce physiological effect. It could be also possible to attach card on a water bottle to transfer 3D wave of the specific substance to water. Such digitized 3D wave could be expressed in any 2D space such as on clothes on walls to affect environment.

4.2 Digitized 3D wave of neurotransmitter effective for brain diseases

Especially digitized 3D wave expressed in 2D space could be effective in for curing brain disease. Brain is protected by blood brain barrier (BBB) which inhibit drug from outside to pass through brain. For example, serotonin deficiency is related to depression. However, there is no way to put serotonin through brain because serotonin could not pass through BBB. Thus, it is needed to bypass BBB.

SSRI (Selective Serotonin Reuptake Inhibitor) class antidepressant drugs are representative bypass drugs and are most widely prescribed currently [31].

They delay the reuptake of serotonin resulting in serotonin persisting longer in brain. Although SSRI class antidepressant drugs are regarded as relatively safe drugs, they have serious side effects such as increasing risk of suicide and increasing tendency for violence [31].

However, if the 3D wave of serotonin itself could be given to patient using digitized 2D card or memorizing ability of water, we can imagine that it could be an ideal safe drug, as there is no substance causing unwanted side effects like SSRI class. Other than serotonin virtually every neurotransmitter which cannot pass through BBB could be medically utilized with the digitization technology shown I this book or using memorizing ability of water. It is being observed that many patients with diseases related with brain such as depression, anxiety, insomnia, dementia, Parkinson's disease and even ADHD (Attention Deficit Hyperactivity Disorder) and autistic children get improved with digitized 2D card containing 3D wave of medically defective neurotransmitter for them [33].

4.3 Human beneficial electricity by UL

Digitized 3D wave could be used to change the characteristic of electricity as well. When digitized 3D wave has been modulated into electricity (UL), harmful nature of electromagnetic wave has been changed in beneficial way to the human. As shown in the result, brain wave is stabilized while watching computer monitor. It is also noticed that the symptoms such as shoulder pain or eye congestion occurred by long time watching computer disappeared [33].

Variety of effects can be expected from electrical devices by application of UL. For example, electrical cooling system does not disturb balance of autonomic nervous system and thus does make people feeling not tired. It is also observed that food made by electrical cooker or microwave tastes better and freshness maintained longer [33].

Especially, mobile phone gives people good feeling and does not get very hot even after long usage. If mobile phone does not get heat, it means that battery can be used longer without losing energy as useless heat. This suggests the possibility of electricity saving by application of UL. It was indeed observed that total electricity usage of a house could be saved 20-40% with application of UL. Further serious researches are expected regarding characteristic of electricity modulated with 3D wave.

It becomes almost common knowledge that electromagnetic wave is harmful to human. However, there is no clear explanation regarding why electromagnetic wave is harmful. The results of this study suggest that electromagnetic waves could be harmful to human because of harmful 3D waves contained in the electricity. This might mean that the current way of generating electricity is not friendly one to human. Our research suggests that it would be possible to generate human beneficial electricity even though intensity of electromagnetic wave remains same.

Electricity exists everywhere and electromagnetic wave is known to disturb autonomic nervous system. If 3D wave of medically effective substance could be modulated into electricity and thus electricity itself could deliver the 3D wave of the substance, it would have limitless possibilities. It could mean that every electrical device could be used as a generator of specific 3D wave.

4.4 Digitized 3D wave functions through holographic space

How could a printed shape in 2D space like common plastic card elicit the same biological reactions as the original substance?

The following explanation could be possible: digitized 3D wave expressed as a certain shape in 2D space is (UN) still related to original substance through hologram, thus characteristic of the original substance still appears even though it is digitized and expressed in plastic card.

In 1961 Eisberg showed by calculation that pilot wave (expressed as 3D wave in this book) is much faster than the speed of light [21, 22]. If 3D wave is faster than the speed of light, it should have negative mass. According to Dirac [32], space is described as a domain of negative energy and negative mass.

It is possible that digitized 3D wave expressed in common plastic card is connecting the 3D wave of the original substance to the space itself (Dirac sea: infinite sea of particle with negative energy) to form the 3D wave of the original substance around the card through holographic space.

Digitized 3D wave could be expressed in various ways in our environment. It would be very nice if we could maintain health by changing environment. Further researches on biological and physiological effects using laboratory experiments and serious clinical studies regarding water memory and digitized 3D wave of the substance are expected.

5. References

[1] Reilly, D. **2001**, The puzzle of homeopathy, *Journal of Alternative and complementary medicine*, 103-109.

[2] Clausius, N., Linde, K., Ramirez, G., Melchart, D., Eitel, F., Hedges, V.L. **1997**, Are the clinical effects of homeopathy placebo effect? Meta. analysis of placebo-controlled trials. *The Lancet*, 350, 834-843.

[3] Dayenase, E., Beauvais, F., Amara, J., Oberbaum, M., Robinzon, B., Miadonna, A., Tedeshit, B., Pomeranz, P.,Fortnerg, P., Belon, J., Saint. Laudy, B., Poitevin, B., Benveniste, J. **1988**, Human basophil degranulation triggered by very dilute antiserum against IgE, *Nature*, 333, 816-818.

[4] Hardi, L., Arnoux, B., Benvenist, J. Effect of dilute histamine on coronary flow of isolated guinea pig heart, *FASEB,* **1991**, 5, A1583.

[5] Benveniste, J., Davenase, E., Ducot, B., Spira, A. Basophil achromasia by dilute ligand, *FASEB*, **1991**, 5, A1008.

[6] Beneniste, J., Arnoux, B., Hardi. L. Highly dilute antigen increase coronary flow of isolated heart from immunized guinea pigs, *FASEB*, **1992**, 5, A1610.

[7] Belon, P., Cumps, J., Ennis, M., Mannaioni, P.F., Roberfroid. M., Sainte. Laudy, J., Wiegan, F.A.C. **2004**, Histamine dilutions modulate basophil activation. *Inflammation Research,* 53, 5, 181-188.

[8] http://www.theguardian.com/science/2001/mar/15/technology2

[9] Aissa, J., Littime, M.H., Attias, E., Benvenste, J. **1993**, Molecular signaling at high dilution or by means of electronic circuitry, *Journal of Immunology,* 150, 146A.

[10] Benveniste, J., Aissa, J., Litime, I., Tsangaris, G., Thomas, Y. **1994,** Transfer of the molecular signal by electronic amplification, *FASEB*, 8. A398.

[11] Aissa, J., Jorgen, P., Litime, I., Behar, I., Benveniste, J. **1995**, Electronic transmission of the cholinergic signal, *FASEB*, 5. A683.

[12] Benveniste, J., Jurgens, P., Aissia. J. **1997**, Digital recording/transmission of the cholinergic signal, *FASEB*, 10. A1479.

[13] Thomas, Y., Schiff, M., Belkadi, L., Jurgens, P., Kahhak, L., Benveniste, J. **2000**, Activation of human neutrophils by electronically transmitted phorbol-myristate acetate, *Medical Hypotheses* 54, 1, 33-39.

[14] Benveniste, J., Aissa, J., Jurgens, P., Hseuh, W. **1997**, Transatlantic transfer of digitized antigen signal by telephone link, *Journal of Allergy and Clinical Immunology*, 99, S175.

[15] Benveniste, J. **1998**, *Digital Biology*: Specificity of the digitized molecular signal. *FASEB*, 10. A1497.

[16] Senekowitsch, F., Endler, P.C., Pongratz, W., Smith, C.W. **1995**, Hormone effects by CD record/ replay, *FASEB*, 5. A2270.

[17] Citro, M., Endler, P.C., Pongratz, W., Vinattieri, C. Smith, C.W., Schulte, J. **1995**, Hormonal effects by electronic transmission. *FASEB*, 5. A2271.

[18] Endler, P.C., Pongraz, W., Wijk, R.V., Waltl, K., Hilgers, H., Brandmaier, R. **1994**, Transmission of hormone information by non-linear means. *FASEB*, 8. A2313.

[19] http://en.wikipedia.org/wiki/Matter_wave

[20] http://en.wikipedia.org/wiki/Pilot_wave

[21] Eisberg, R. M. **1961**, Fundamentals of Modern Physics, John Wiley and Sons, Inc., New York, N.Y.) pp. 140-146

[22] Tiller, W. Science and Human Transformation, *PAVIOR*, **1997**, www.tiller.org.

[23] Kim, H. W. Apparatus for delivering substance information, *Korean patent*, **2010**, 10- 2010- 0012200.

[24] Kim, H. W. Characteristic and application of apparatus for transcribing 3D wave. *Journal of Applied Subtle Energy,* **2011**, 9, 2, 32-41.

[25] Kim, H. W. New approach controlling cancer: water memory. *Journal of Vortex Science and Technolgy*, **2013**, 1, 1-8.

[26] Montagnier, L, Aissa, J, Del Giudice, E., Lavallee, C., Tedshi, A., Vitiello, G. DAN wave and water. **2011,** *J. Phys.* / Conf. Ser. 306 (http://arxiv.org/pdf/1012.5166)

[27] Oh, H. K., Oh, Y., Oh, J. Measuring and Characterization of rotational electromagnetic waves, *Journal of Applied Subtle Energy,* **2007**, 5, 1, 24-29.

[28] Oh, H. K. Method and apparatus for measuring circularly polarized rotating electromagnetic wave using magnetic field, *Korean patent,* **2010**, 10-0631869.

[29] Yoo, S. K., Lim, M. Y., Oh, S. M., Yoo, K. B., Shin, Y. H., P, P., Lee, W. A study to verify the confidence-degree of the analyzing method using BRS as a kind of bio-information analysis method. *Journal of the Korean Jungshin Science Society*, **1998**, 2, 2, 51-53.

[30] Kim, H.W. Life of Water: A Cure for Our Body, *Bookscom*, **2008**

[31] http://en.wikipedia.org/wiki/Serotonin

[32] http://en.wikipedia.org/wiki/Dirac

[33] http://cafe.daum.net/khwsupport

i want morebooks!

Buy your books fast and straightforward online - at one of world's fastest growing online book stores! Environmentally sound due to Print-on-Demand technologies.

Buy your books online at
www.get-morebooks.com

Kaufen Sie Ihre Bücher schnell und unkompliziert online – auf einer der am schnellsten wachsenden Buchhandelsplattformen weltweit! Dank Print-On-Demand umwelt- und ressourcenschonend produziert.

Bücher schneller online kaufen
www.morebooks.de

 VDM Verlagsservice-
gesellschaft mbH

VDM Verlagsservicegesellschaft mbH
Heinrich-Böcking-Str. 6-8   Telefon: +49 681 3720 174   info@vdm-vsg.de
D - 66121 Saarbrücken   Telefax: +49 681 3720 1749   www.vdm-vsg.de

MIX
Papier aus verantwortungsvollen Quellen
Paper from responsible sources
FSC® C105338

FSC
www.fsc.org

Printed by Books on Demand GmbH, Norderstedt / Germany